Crystal Research and Technology

5

Journal of Experimental and Industrial Crystallography

Zeitschrift für experimentelle und technische Kristallographie

Volume 23
1988
Number 5 May

Akademie-Verlag · Berlin
VCH Publishers, Inc. · New York/N.Y.

ISSN 0232-1300 Cryst. Res. Technol., Berlin **23** (1988) 5, 583–712, K81–K92

Directions to the author

In the journal "Crystal Research and Technology — Kristall und Technik" as a rule the following categories of papers will be printed: original papers, review articles, short notes, and book reviews, preferentially in English. Complete information necessary for type-setting and a "Short Title" should be added.

The manuscript should not exceed 10 pages and be type-written on one side only in the format A 4 (21 × 29,7 cm) on white ink-proof paper, with a two-lined distance and a left-hand free edge of 3 cm, and submitted in 2 copies. All sheets of the paper belonging to the contribution are continuously to be numbered.

The head of the manuscript should be arranged in this manner: initials and name of the author; institution, and title of the paper. Then two abstracts should follow, one in the original language and the second in German, English, French, or Russian, respectively.

Formulae, signs and symbols, letters etc. being wanting on the type-writer, should be distinctly written with penholder. O as zero is to be underlined, the capital O is not. Greek letters should be underlined with red ink.

Paragraphs should be marked by indenting of at least 3 touches.

Concerning the marking of the manuscript should be noted:
headings should be underlined, sub-headings not, and indented by 9 touches. The titles of tables are not to be underlined. Other markings in the text will be done by the editorial staff. They mean:

light-red ～～～～～～ = small capitals
blue ～～～～～～ = in italics

light-red | perpendicular straight red line on the left edge = small printing

Tables are to be added on separate sheets of paper at the end of the contribution.

Foot notes should be continuously written on a separate sheet of paper.

Figures (Fig.) — drawings and photographs — should be not more than 8 and submitted (2 copies) in a special envelope: photographs as white high-gloss prints (inscription on a transparent covering sheet) and drawings as distinct sketches. On the back of the submitted figures the reduction scale in tenth (9/10, 8/10, 5/10 e.g.), author's name, short title of the paper, and the number of the figures should be given.

Captions for the figures should be submitted on a separate sheet of paper.

References should be written without numbers in alphabetical order of the names at the end of the contribution. Cited names in the text stand for the references. The references should contain the following items, for example:

for books: RAST, K.: Physikalisch-chemische Rechnungen, Berlin 1958
for journals: KRAUSE, H. N., MÜLLER, K.: Cryst. Res. Technol. 16 (1981), 70

Add, please, the short title, running title, about 6 keywords, and the academic grades of the authors.
The authors will be sent two galley proofs of review articles and original papers. Proof reading of short notes and book reviews will be done by the editorial staff.

For each of these papers a total of 50 reprints is available free of charge. Requests for additional maximum 200 reprints (against payment) should be given together with the submitted manuscript.

Founded 1966 as
Kristall und Technik

Crystal Research and Technology 5

VOLUME 23 · 1988 · NUMBER 5

Journal of Experimental and Industrial Crystallography

Akademie-Verlag · Berlin
VCH Publishers, Inc. · New York/N. Y.

For the USA and Canada:
Crystal Research and Technology, ISSN 0232-1300, is published monthly by VCH Publishers, Inc., 303 NW 12th Avenue, Deerfield Beach, FL 33442-1788; Telex 5101011104. VCH Pub. Telefax (305) 428-8201; Telephone (305) 428-5566 or 1-800-422-8824 (toll free).

Second class postage paid at Deerfield Beach, FL 33441.

Annual subscription price: US $ 325.00 including postage and handling charges.

U.S. Postmaster: Send address changes to VCH Publishers, Inc., 303 NW 12th Avenue, Deerfield Beach, FL 33442-1788.

Inquiries, subscriptions, and orders for single copies from customers in the USA and Canada should be directed to VCH Publishers, Inc., 303 NW 12th Avenue, Deerfield Beach, FL 33442-1788, USA;
from all other countries to
Akademie-Verlag Berlin, Leipziger Str. 3 – 4, DDR-1086 Berlin.

Papers should be directed to the editor's office, Scharnhorststraße 20, DDR-7030 Leipzig, or to the Editor-in-Chief Prof. Hermann Neels, Zeititzer Weg 25, DDR-7254 Machern,
or to one of the members of the Editorial Board or Advisory Board.

Journal „Crystal Research and Technology"

Herausgeber: Im Auftrage des Akademie-Verlages Berlin herausgegeben von Prof. Dr. Hermann Neels.

Verlag: Akademie-Verlag Berlin, Leipziger Str. 3 – 4, DDR-1086 Berlin; Fernruf: 2236221 oder 2236229; Telex-Nr.: 114420; Bank: Staatsbank der DDR, Berlin, Kto-Nr.: 6836-26-20712.

Chefredakteur: Prof. Dr. Hermann Neels, Zeititzer Weg 25, DDR-7254 Machern; Fernruf: Brandis 3442.

Anschrift der Redaktion: Scharnhorststraße 20, DDR-7030 Leipzig; Fernruf: Leipzig 310502.

Veröffentlicht unter der Lizenznummer 1526 des Presseamtes beim Vorsitzenden des Ministerrates der Deutschen Demokratischen Republik.

Gesamtherstellung: VEB Druckerei „Thomas Müntzer", DDR-5820 Bad Langensalza.

Erscheinungsweise: Das Journal „Crystal Research and Technology" erscheint monatlich. Die 12 Hefte eines Jahres bilden einen Band.

Bestellnummer dieses Heftes: 1082/23/5.

Cryst. Res. Technol.	23	1988	5	585–594

Original Papers

Chr. N. NANEV

Institute of Physical Chemistry, Bulgarian Academy of Sciences
Sofia, Bulgaria

On the Morphological Instability of Growing Crystals (I)

Morphological Peculiarities of the Transition Shapes of Crystals in Diffusion-controlled Regime of Growth

The shape and size of the gross morphological defect which occurs (in the form of a depression) as a result of the diffusion inhomogeneity in the matter supply on the growing crystal face with supercritical dimensions are investigated. The transition shapes of similar skeletal crystal faces are determined by considering the configuration of the steps which are moving (and accumulating) from the edge toward the centre of the crystal face. The basic parameter which is evaluated, is the width of the macroscopically flat periphery which grows around the depression and its alteration when the crystal is enlarged. The theoretical analysis is carried out by presuming that no transportation of crystallizing material occurs parallel to the crystal face. The calculations are based on the different thicknesses of the Nernst diffusion layer, a magnitude which can be easily determined experimentally by some interferometric technique, over the various regions of the growing face.

Es werden die Beschaffenheit und die Ausdehnung des groben morphologischen Defekts untersucht, welcher infolge der diffusions-bedingten Übersättigungsinhomogenität auf den Kristallflächen überkritischer Größe (in der Form einer Vertiefung) entsteht. Die Übergangsformen solcher skelettartigen Kristallflächen werden auf Grund der Stufenkonfigurationen bestimmt, welche bei der Bewegung (und Anhäufung) der auf den Kristallecken entstandenen und sich in Richtung Flächenmitte bewegenden Wachstumsstufen auftreten. Die Breite der um der Vertiefung wachsenden makroskopisch ebenen Peripherie und ihre Veränderung mit der Vergrößerung des Kristalls ist der berechnete Grundparameter. Die theoretische Analyse wurde unter der Voraussetzung durchgeführt, daß kein Transport des zu kristallisierenden Stoffes parallel der Kristallfläche stattfindet. Die Berechnungen beruhen auf dem Unterschied in der Dicke der Nernst'schen Diffusionsschicht über die verschiedenen Bereiche der wachsenden Kristallfläche — eine experimentell (interferometrisch) leicht meßbare Größe.

1. Introduction

The morphological stability of crystals is a long standing focus of special interest, attracting the attention during three of the initial international crystal growth conferences held in Boston, Birmingham and Tokyo (PEISER; FRANK et al.; JACKSON et al.). Since the classical investigations by MULLINS and SEKERKA, the problem concerning the stability during an isotropic surface kinetics can be considered as being solved, at least in principle. The quantitative terms defining the critical dimensions, which when surpassed lead to the disappearing of the polyhedral shape of the crystal have been derived also (CHERNOV; CAHN). The transition to skeletal growth, i.e. after the morphological stability has disappeared, on the other hand, has been relatively less investigated theoretically. However, quite many systematic experimental observations exist (zinc and cadmium in inert gas media (NANEV, IWANOV), ice (NENOW et al.), NaCl and KCl (VETTER et al.) and KDP (STAJNOVA, NANEV) from aqueous solutions, electrodeposition of silver (RASHKOV, NANEV) concerning the critical size and the growth of skeletal crystal faces under diffusion control conditions. It must be noted that the shape of the depression is always "dish-like" and no "compensation profile" (NANEV 1987) is present.

The present study is aimed at the explanation of the shape and dimensions of the gross morphological defect — the depression which occurs upon the hitherto macroscopically flat face during its growth beyond the critical size. It appears that the easiest way to achieve this goal is by the determination of the macroscopically flat periphery, which continues to grow around the depression. In fact it determines the growth pattern, since the region of the depression itself now grows under diffusion control conditions, i.e. substantially slower and therefore retains to a higher degree the shape, acquired as a result of the consecutive positions of the inner edge of the macroscopically flat periphery. The present investigation determines: 1. The diameter of the depression at each moment of the crystal growth; 2. Its chape (polygonized or round) and 3. The qualitative character of the profile (cross-section) of the depression. The approach is two-dimensional, using cross-sections[1]).

2. Basic assumption

The growing crystal consumes material from the vicinity, which in the case of stagnant solutions is supplied by diffusion only. Its form has been determined through the competition between the growth kinetics and diffusion. The case of anisotropic surface kinetics which is treated in the present study differs substantially from that of an isotropic surface kinetics, which characterizes the growth of crystal spheres. However, we will compare the two different cases in an effort to formulate the problem in a clear-cut way.

In the case of a single crystal sphere with radius R, MULLINS and SEKERKA use the phase boundary condition:

$$(C - c_s)\frac{dR}{dt} = -D\left(\frac{\partial c}{\partial r}\right)_{r=R} \tag{1}$$

where C is the concentration of the precipitate, c_s is the solution concentration on the crystal surface. dR/dt is the growth rate of the crystal, D and dc/dr are, respectively, the diffusivity and concentration gradient (with respect to the distance r in spherical coordinates) in the solution.

In the literature one can encounter most frequently the radial distribution of the concentration c in the solution around the spherical crystal, which is derived from the solution of the Laplace equation (MULLINS, SEKERKA; CAHN).

$$c - c_\infty = \frac{(c_s - c_\infty)\,R}{r} \tag{2}$$

where c_∞ is the bulk concentration in the solution (at $r = \infty$). Strictly spoken, it is valid for diluted solutions and in the immediate vicinity of the crystal surface (since

[1]) This is simpler, since when we consider a cross-section along a major crystallographic direction, the elementary growth steps do not change their orientation. Therefore, their growth rate depends on the local supersaturation only. When we consider a second cross-section of a characteristical crystallographic angle, we can determine also the point when the periphery of the depression starts to polygonize.

it gives a too large concentration depletion at larger distances and does not concerve the mass in a diffusion problem (CAHN). A better approximation in the vicinity of the crystal is provided by the more accurate solution of the type:

$$[(c_s - c_\infty) R/r] \exp(-r^2/4Dt) .$$

In this case the concentration gradient on the surface of the crystal is presented by:

$$\left(\frac{\partial c}{\partial r}\right)_{r=R} = \frac{c_\infty - c_s}{R} . \tag{3}$$

In the following considerations we will treat only the case of anisotropic surface kinetics. After the critical size l_c[2]) has been reached, the flow of crystallizing material, obviously, will no longer be identical towards all sites on the crystal face. The fact that a depression is formed on the face is alone an evidence for this conclusion — there the flow is smaller since diffusion transportation mode prevails. But, on the other hand, we do not observe an abrupt disappearing of the flat shape of the entire face. A macroscopically flat region remains growing along the periphery, where still the stabilizing factor of the anisotropic surface kinetics is predominant, i.e. the compensating mechanism still works (CHERNOV; NANEV 1987). However, this flat rim is altered, it becomes narrower. Obviously, this is a result of the continuing competition between the diffusion and kinetic factors. One can see that within the framework of the same crystal face there are regions which simultaneously grow under diffusion and kinetic control, i.e. the latter is not completely surmounted. This fact provides evidence to accept that even within the framework of one and the same crystal face various mass conservation boundary conditions may be valid. Equation (1) obviously represents the principle of the law of mass conservation (see also PAPAPETROU). Therefore, when we use the respective growth rate V for the face of the crystal polyhedron, i.e. V instead of dR/dt, in this case also equations similar to eq. (1) should be valid. By presuming that no transportation of crystallizing material occurs parallel to the crystal face, we will accept that it is valid in a detailed manner, i.e. along the entire face length. On the basis of this sole presumption, this paper will describe the growth of the skeletal forms of the crystal faces.

The most substantial approximation in the present work is the presumption that

$$\frac{\partial c}{\partial r} \approx \frac{\Delta c}{\Delta r} = \frac{c_s - c_\infty}{\delta_N(l)} , \tag{4}$$

where δ_N is the thickness of the Nernst diffusion layer, and as a matter of principle, it may be accepted that δ_N may be a function of the crystal size l. In addition it is very substantial to mention that this relationship is valid for concentrated solutions also.

The resemblance between the two expressions (eqs. (3) and (4)) is obvious. They become identical at $R = \delta_N$, i.e. in the case of crystal with dimensions l within the approximate range of 20 to 250 μm. In fact those are the sizes when usually the diffusion phenomena considered here appear, or let us put it in another way — the average value of the concentration gradient according to Nernst's conceptions is equal to the "modulus" of the concentration gradient, calculated for the surface of the spherical crystal on the basis of the Laplace equation. Therefore, the use of δ_N for such (or nearly similar) dimensions is well motivated and has the advantage to reflect the different availability to diffusion of the edges and centres of the crystal faces.

---

[2]) In this case the absolute size of l_c is of no importance. The critical dimension of the crystal is used only as a starting point in the considerations.

The present investigation is based on the latter circumstance, i.e. the different thickness of Nernst's diffusion layer over various regions of the crystal face.

According to the theory of BURTON et al. the close packed faces of real crystals grow by a screw-dislocation mechanism. The growth rate is a linear function of the supersaturation at relatively high supersaturation values and quadratic at rather low values. We will consider them as follows: 1. The kinetic coefficient β for the respective crystal face determines its growth rate V at high supersaturations:

$$V = \beta(c_s - c_e)/c_e\,, \tag{5}$$

where c_e is the equilibrium concentration in the solution. Thus, it follows from equations (1), (4), and (5)

$$\frac{c_s - c_e}{c_\infty - c_e} = \frac{1}{1 + \delta_N\beta(C - c_s)/Dc_e}\,. \tag{6}$$

2. At low supersaturations and a screw -dislocation growth mechanism

$$V = b(c_s - c_e)^2/c_e^2 \tag{7}$$

where b is the proportionality coefficient.

The solution of the quadratic equation for $(c_s - c_e)$ leads in this case to the following expression

$$c_s - c_e = \frac{Dc_e^2}{2b\delta_N(C - c_s)}\left[-1 + \sqrt{1 + \frac{4b}{Dc_e^2}(C - c_s)\,\delta_N(c_\infty - c_e)}\right]. \tag{8}$$

A spin-off of this expression is that it provides a simple quantitative criterion for the limit up to which the crystal can grow in a kinetic regime. Obviously, the condition is $c_s = c_\infty$. It is well known from the rules for approximate calculations that this condition can be achieved with different accuracy. Thus, in order to obtain $c_s = c_\infty$ with an accuracy of 0.1%, the term in the root of equation (8) must be less than or equal to 1.093. But if we require an accuracy of only 1%, then the value of the term in the root should be less than or equal to 1.328. These conditions lead directly to the equation:

$$\frac{D}{b\delta_N} \geqq N_i C(c_\infty - c_e)/c_e^2\,, \tag{9}$$

since generally $C \gg c_s$ and $N_1 = 40$ at an accuracy of 0.1% or $N_2 = 13$ at an accuracy of 1%, and so on. The criterion (eq. (9)) comprises experimentally accessible magnitudes, $-\,b$ and δ_N.

As mentioned in the introduction (see foot note 1), when we apply a two-dimensional approach, the role of the local deviation from equilibrium consists only in the fact that it determines the lateral velocity v_l of spreading of steps, which pass through the given point on the crystal face (CAHN). Since screw dislocations are emerging almost everywhere on the crystal face, the sites of generation of new steps are determined by the highest supersaturation. In the case of diffusion growth conditions those are the regions near the crystal edges. In fact we have observed the initiation of steps along the periphery of the face and their shift toward the centre in the case of zinc (NANEV, IWANOV) and by the electrocrystallization of silver (RASHKOV, NANEV), by applying in the latter case of a Nomarski interference contrast (Fig. 1). On reaching the face centre the steps are either annihilated, or if the face is larger, and consequently the inhomogeneity of the supersaturation is also substantial, they slow down their movement and a depression is formed (Fig. 2).

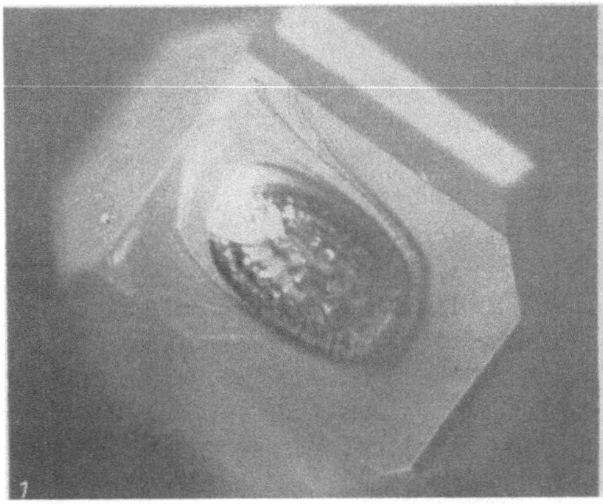

Fig. 1. Silver single crystal electrolytically grown. The depression in the centre of the cubic face is due to the diffusion limited supply (c.d. 450 mA/cm²). The Nomarski interference contrast visualizes fine steps, spreading from one end of the face, as a result of the specific disposition of the screw dislocation sources and a certain asymmetry of the anode

Fig. 2. Depression on the prismatic face of a KDP single crystal, grown from an aqueous solution during cooling. A tendency is observed to retain the width of the macroscopically flat periphery almost identical in all crystallographic directions

3. Diameter of the depression

If d stands for the step density and h is the step height, the following relationship (CAHN) is obvious:

$$V = v_1 dh \,. \tag{10}$$

Thus, the condition for a conservative shift, i.e. without a major shape change, of the entire face may be written as follows: $V =$ const, leading locally, for each site to the expression

$$\frac{v_1}{v_1'} = \frac{d'}{d} \,. \tag{11}$$

Now, let us agree to denominate the centre of the crystal face by the index "0", the edges — by "1" and the momentary position of the inner edge of the macroscopically flat periphery by "k". For instance, the moment when the depression is initiated in the centre of the face with critical dimensions l_c is characterized by the critical density of the steps d^{cr}, i.e. $d^{cr} = d^0$ at $l = l_c$. The value of d^{cr} is automatically established when a certain "critical" lagging in the movement of the steps is reached, resulting in their piling. This is due to mutual hindrance — e.g. overlapping of their diffusion fields. On the other hand, as shown by NANEV (1987), the position of the inner edge of the macroscopically flat periphery, which determines its width, will always be characterized by the point where this d^{cr} value is reached, respectively β^{cr} and b^{cr}.

At this point, as a matter of principle, we must differ from each other two possibilities (BURTON, CABRERA, FRANK):

1. At high supersaturations, v_1 is independent on its value.
2. At low supersaturations a linear relationship $v_1 = \alpha(c_s - c_e)/c_e$ exists, where α is the respective kinetic coefficient of the treated case.

Obviously, the second possibility is more important. In fact, even if the supersaturation in the bulk $(c_\infty - c_e)/c_e$ is high, it diminishes on the crystal surface, due to diffusion limitations. Even the mere fact that a depression is formed, is an evidence for this statement since one has to keep in mind that the shift of steps in its vicinity is considered. Then for "rough" steps we can write:

$$\frac{d^1}{d^k} = \frac{\alpha^k(c_s^k - c_e)}{\alpha^1(c_s^1 - c_e)}, \tag{12}$$

where α^k is the kinetic coefficient which characterizes the rate at which new crystal layers extend laterally past the inner edge of the macroscopically flat periphery (in the case of a single component system $\alpha^k = \alpha$). In a first approximation d^1 can be considered also constant[3]).

If we use equations (6) and (8) for the local supersaturation applying them in a detailed manner, and since generally $C \gg c_s$, we may write:

$$\frac{c_s^k - c_e}{c_s^1 - c_e} = \frac{c_s^{0,\mathrm{cr}} - c_e}{c_s^{1,\mathrm{cr}} - c_e} = \mathrm{const} = \frac{1 + \delta_N^1 \dfrac{\beta^1 C}{D c_e}}{1 + \delta_N^k \dfrac{\beta^k C}{D c_e}}, \tag{13}$$

$$\frac{c_s^k - c_e}{c_s^1 - c_e} = \frac{c_s^{0,\mathrm{cr}} - c}{c_s^{1,\mathrm{cr}} - c_e} = \mathrm{const} = \frac{\delta_N^1 b^1 \left[-1 + \sqrt{1 + 4\dfrac{b^k \delta_N^k C(c_\infty - c_e)}{D c_e^2}}\right]}{\delta_N^k b^k \left[-1 + \sqrt{1 + 4\dfrac{b^1 \delta_N^1 C(c_\infty - c_e)}{D c_e^2}}\right]}. \tag{14}$$

In the case of a single component system, obviously during the enlargement of the crystal in the right side of expressions (13) and (14) only δ_N^1 and δ_N^k may change. On the other hand, WILCOX has calculated and we have observed (NANEV, DYULGEROV) that the isoconcentrates around a growing cube display a spherical (respectively in two-dimensions — circular) symmetry with a radius of curvature approximating that of the inscripted sphere, the effect being especially pronounced in the case of the isoconcentrate lines situated more outwards. The boundary of the Nernst diffusion layer must be perceived as the most outward positioned isoconcentrate around the growing crystal.

Let us first accept that during the enlargement of the crystal within certain limits δ_N retains its radius of curvature[4]). Thus we can calculate accurately the width f of the macroscopically flat periphery when the size of the crystal l is enlarged. Latter on, we will see what happens if this first approximation is not valid and the radius of curvature changes: a. Obviously one of the possibilities is that δ_N^1 does not change in both equations (13) and (14) and then[5])

$$\delta_N^k = \delta_N^{0,\mathrm{cr}} = \frac{l_c}{2} \operatorname{tg}\left(\frac{1}{2} \arcsin \frac{l_c}{2\varrho}\right) + \delta_N^1 = \mathrm{const}\,. \tag{15}$$

---

[3]) But in case that step sources with different activity — e.g. pairs of screw dislocations of opposite sign, etc. (NANEV 1977) become active in the region around the crystal edge d^1 may change its value and eventually alter the width f of the macroscopically flat periphery.

[4]) Naturally, in this case no spherical isoconcentrates are formed around the growing crystal.

[5]) Follows directly from the expression for the "arrow" of a circular segment.

Here ϱ is the radius of the "spheric" isoconcentrate line over the crystal surface. The geometric shape of the crystal (Fig. 3a) is built according to this assumption. In brief the procedure is as follows.

Over the faces of squares which increase in size we draw arcs with constant radius of curvature ϱ[4]), which are elevated upon the peaks of squares at equal distances δ_N^1. These arcs approximate the effective thicknesses of the Nernst's diffusion layer over the different regions of the crystal face. The distance $\delta_N^{0,cr}$ over the centre of the square

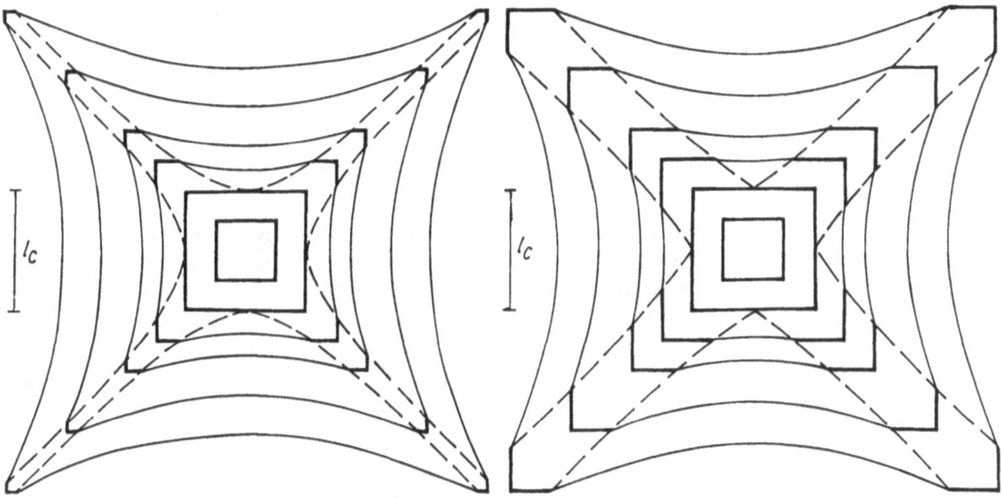

Fig. 3. Sectional view along {100} of six consecutive stages in the enlargement of a cubic crystal: The first is in a kinetic regime – dimensions less than l_c, the second – just by l_c, whereas the next four are in a diffusion regime (size surpassing l_c); details given in the text: a) $\varrho = 2.7 l_c$, b) $\varrho \cong \frac{\sqrt{2}}{2}(l + \varrho_N^1)$

with critical size l_c is the parameter, whose constant value determines latter on the positions of the inner edge of the macroscopically flat peripheries, the so-called "kerb-points" (NANEV 1987) – during the following enlargements stages of the crystal. The flat peripheries around the edges of the larger crystals are presented in Figure 3a with solid lines. The trajectories traced by the inner edge point (kerb-point) shown in Figure 3a with dashed lines are to a great extent independent of the curvature of the Nernst isoconcentrational lines. The depression region is purposely drawn heavily exaggerated.

Figure 3a offers a visual pattern of the skeletal and dendritic growth modes. Obviously, their growth is enhanced by adsorption phenomena. The latter is stronger on the depressed parts of the "faces". The stronger the adsorption phenomena, the more growth rate slows down at the site and the crystal acquires the shape of a dendrite. In the reverse case, when the crystal grows with an isometric form, adsorption phenomena upon the smooth (close packed) crystal faces may be altogether ineffective.

b) The second possibility, resulting from equation (13), is a linear relationship between δ_N^k and δ_N^1:

$$\delta_N^k = A + B\delta_N^1 , \tag{16}$$

where A and B are constants.

c) The change of δ_N^k as a result of the alteration of δ_N^1 is expressed, according to equation (14) by a rather complicated function. However, since the magnitudes are changed under the square root only slightly, we can accept them in a first approximation as constants. Thus, a linear relationship is likely to exist between δ_N^k and δ_N^1.

In both cases b) and c), depending on δ_N^1 and B, the flat periphery will remain wider or will become narrower faster than shown in Figure 3a. Within the framework of some limits the general character of the dependence of f on the size l of the crystal is retained. Moreover, the width f of the macroscopically flat periphery, shown in Figure 3a, resembles closely the one reported earlier (NANEV 1987). In fact, within the range $1.1 \leqq l/l_c < 2.0$ it is described approximately by the same relationship

$$\frac{f}{l_c} = \frac{l}{l_c} - 1.3 \text{ Arc cosh} \left(\frac{l}{l_c}\right). \tag{17}$$

At higher l/l_c values, f/l_c begins to decrease slightly and practically follows a linear relationship.

It must be especially stressed that in fact during the enlargement of the crystal, the radius of curvature of Nernst's diffusion layer also increases. Simple geometrical structures show that if the radius ϱ (which indicates the position of δ_N) decreases during the enlargement of the crystal, then the width f of the macroscopically flat periphery will also decrease, and vice versa, if the radius augments — the width of the macroscopically flat periphery will also increase. Figure 3b represents a graphic display of a crystal (drawn by a procedure which is closely similar to the just described one) with δ_N^k and a radius ϱ of the spheric isoconcentrates $\varrho \cong \frac{l}{2 \sin (\theta/2)} + \delta_N^1 \cos (\theta/2)$ and $\delta_N^1 = 1/\sqrt{l}$, where θ is the central angle (90° in the present case).

The width w of the depression is determined at any moment during the enlargement of the crystal by the simple expression $w = l - 2f$.

4. The shape of the depression

The shape of the depression is determined in a convinient way, as already mentioned (see foot note 1), by considering other cross-sections[6]) also — in the simplest case in other close packed directions. It can be immediately noted that if we consider the diffusion supply only, the periphery of the depression will be always circular, since the cross-section of the isoconcentrational spheres with flat planes are circles. Therefore it is clear that in the case of variable radii of the isoconcentrational spheres, the shape of the depression will also remain circular.

However, experiments provide evidence that the periphery of the depression is often polygonized. This is a result of the effect exerted by the kinetic factor. In this case the principle of correspondence interferes — the depression is polygonized "reversely" to the large crystal (Fig. 4). The phenomenon is modified by other factors too, e.g. adsorption and inclusion of impurities. Under these circumstances a "parallel" orientation of the depression along the faces of the crystal (Fig. 5) is obtained as a result of the reversal of the growth rates. It must be mentioned that in the latter case a minimum width of the macroscopically flat periphery is observed. Under heavy passivation conditions a macroscopical bunching effect, i.e. a terraced

[6]) In this case, naturally, the kinetic coefficient α, i.e. the spreading rate of the elementary steps, is changed.

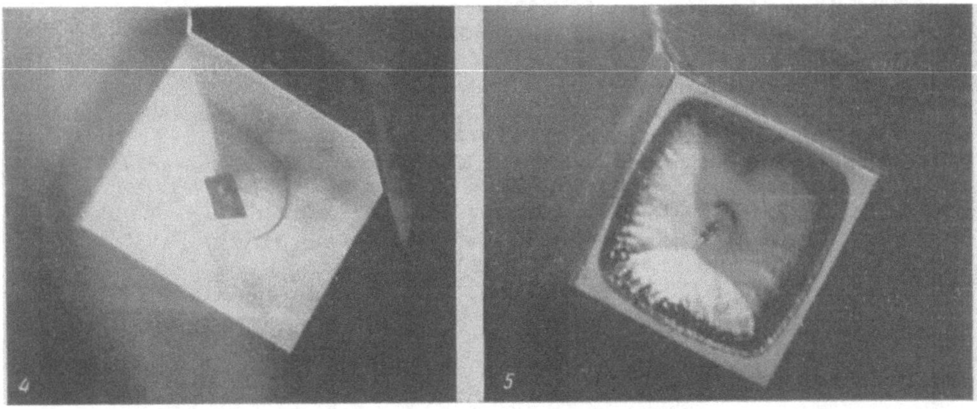

Fig. 4. Cubic face of a silver single crystal grown electrolytically. A nearly rectangular depression is formed according to the principle of correspondence (i.e. in a "reversed" orientation)

Fig. 5. Silver single crystal with parallel orientation of the depression and minimum macroscopically flat periphery. Distinct growth inhibition symptoms were observed

depression is observed. In some cases, if the passivation is very strong, even a shift of the centre of the depression occurs.

5. Cross-section of the depression

It is an experimentally established fact that the shape of the depression profile is not of the "compensation" type (NANEV 1987; CHERNOV), but rather resembles a dish (NANEV, IWANOV; NENON et al.; YUMOTO et al.; STAJNOVA, NANEV; RASHKOV, NANEV). A possible explanation of this fact is that the shape is designed by the consecutive positions of the inner edge point of the macroscopically flat periphery during the enlargement of the crystal above l_c. The compensation profile, in a rather reverse way, traces the shape of faces with subcritical dimensions only. If the critical size l_c is surpassed, the compensation mechanism no longer acts in a global manner, i.e. covering the entire face. Instead it affects the region of the macroscopically flat periphery only and consequently can not determine the shape of the depression since at that very site it is inactive.

This project has been completed with the financial support of the Committee for Science at the Council of Ministers under contract No KH 168/1987.

References

BURTON, W., CABRERA, N., FRANK, F. C.: Phil. Trans. **A 243** (1951) 299
CAHN, J. W.: in: Crystal Growth, Oxford 1967, p. 681
CHERNOV, A. A.: Kristallografiya 8 (1963) 87
FRANK, F. C., MULLIN, J. B., PEISER, H. S.: Proc. of the Second ICCG, Birmingham 1968 (J. Cryst. Growth **3/4** (1968))
JACKSON, K. A., KATO, N., MULLIN, J. B.: Proc. of the Fourth ICCG, Tokyo 1974 (J. Cryst. Growth **24/25** (1974))
MULLINS, W. W., SEKERKA, R. F.: J. Appl. Physics **34** (1963) 323
MULLINS, W. W., SEKERKA, R. F.: J. Appl. Physics **35** (1964) 444

NANEV, Chr.: Cryst. Res. Technol. **12** (1977) 587
NANEV, Chr.: Cryst. Res. Technol. **22** (1987) 157
NANEV, Chr., DYULGEROV, A.: Cryst. Res. Technol. **17** (1982) 19
NANEV, Chr., IWANOV, D.: Cryst. Res. Technol. **17** (1982) 575
NENOW, D., STOYANOVA, V., GENADIEV, N.: J. Cryst. Growth **66** (1984) 489
PAPAPETROU, A.: Z. Kristallographie (A) **92** (1935) 89
PEISER, H. S.: Crystal Growth (Proc. of 1-st ICCG, Boston 1966, Oxford etc. 1967
RASHKOV, R., NANEV, Chr.: to be published
STAJNOVA, M., NANEV, Chr.: to be published
VETTER, J., WINZER, A., EMONS, H.-H.: Kristall und Technik **11** (1976) 163
WILCOX, W. R.: J. Cryst. Growth **37** (1977) 229
YUMOTO, H., HASIGUTI, R. R., KANEKO, T.: J. Cryst. Growth **75** (1986) 289

(Received, accepted October 14, 1987)

Author's address:

Dr. Christo N. NANEV
Institute of Physical Chemistry, Bulgarian Academy of Sciences
akad. G. Bonchev bl. 11
1040 Sofia, Bulgaria

Cryst. Res. Technol.	23	1988	5	595–603

N. N. SIROTA, T. B. BIZHIGHITOV

MGMI, Department of Physics, Moscow

Polymorphic Transitions of Ice at a Pressure up to 2500 MPa in the Temperature Range 90—300 K

In honour of Professor HERMANN NEELS' 75th birthday

In the present work the phase $P-T$ diagram of ice is constructed in the temperature range 90–250 K and is ascertained at temperatures 250–300 K and pressures up to 2500 MPa. A new modification of ice – ice XII – was found. Compressibility and Grüneisen coefficients for a number of ice modifications were determined.

1. Introduction

Phase $P-T$ diagram of ice, presently known, has been constructed predominantly according to the data of Tamman and Bridgman concerning mainly equilibrium water – ice lines and the existence of modifications at temperatures exceeding 230 K.

The supposition of POLIAN and GRIMSDITCH on the existence of ice X modification has not been confirmed by LIN-GUN-LIU. In 1985 ice XI modification was discovered at temperature below 77 K and pressures up to 150 MPa. We studied the temperature and pressure region which has been practically unexplored earlier.

2. Experimental part

We studied the change of specific volume of different ice modifications caused by temperature and pressure variations, including phase transitions measurements were taken in isothermal conditions in case of growing and decreasing pressure and also in isobaric conditions when temperature grew and decreased.

The measurements of volume change were realized step by step, both in isothermal and in isobaric conditions. At each step the measurements had not been taken until equilibrium state was achieved.

An ice sample of twice distilled water (a cylinder with diametre 12 mm and 15 mm high) was placed into cylindric ampoulla of Teflon with the walls 0.5 mm thick. The ampoulla with the sample was placed into a high-pressure chamber cooled by liquid nitrogen or its vapour.

The measurements of ice sample volume variation were realized within the "cylinder-piston" scheme, according to the displacement of the punch (piston) measured by a specially designed reversive transducer. Elastic deformation of punch and chamber was taken into account.

The pressure in the measuring cell was determined according to the variation of the resistance of gallium and tin wires placed within the cell and the readings of the pressure transducer "Sapphire", connected in series with the press manometer ((D0137A). Precision of pressure measurements was 0.4%; that of the volume variation measurements was 0.15%.

A certain temperature in the high-pressure chamber was sustained with the help of cooling nitrogen vapour, its temperature having been regulated by special heaters.

The temperature was controlled by a copper-constantan thermocouple placed in the measuring cell. In isothermal measurements the temperature variations did not exceed ± 2 K.

3. Results

As an example, in Figure 1 isotherms of relative volume variation $\Delta V/V_0$ of the sample versus pressure are shown for different temperatures. Ice sample was placed into the chamber at 250 K and was cooled to required temperature at constant pressure

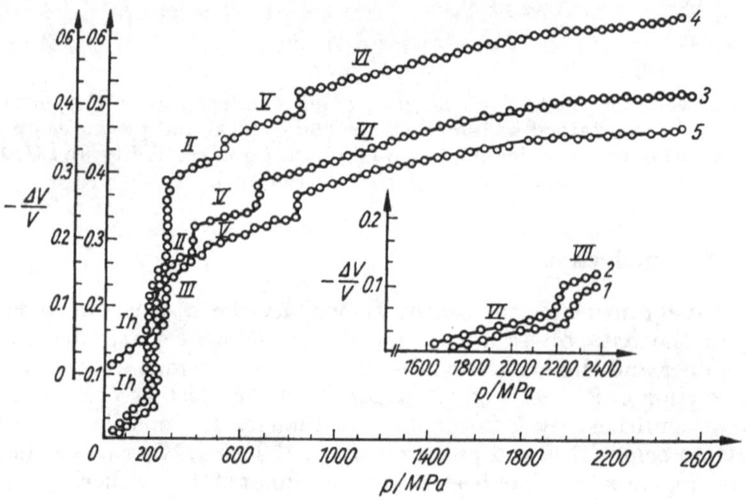

Fig. 1. Isotherms of relative volume variation of ice sample for growing pressure at different temperatures. Curves:
1: 315 K, 2: 300 K, 3: 253 K, 4: 213 K, 5: 233 K

(10 MPa). Then, the temperature being constant, the pressure was gradually increased, with a delay after each 200 MPa increase. Sample volume was measured at the end of each delay. When the pressure reached 2300–2500 MPa, the decrease of the pressure was carried out similarly.

In the isotherms of Figure 1 phase transitions are distinctly manifested.

In case of isobaric measurements the ice sample, as in isothermal case, was placed into the chamber at 250 K. Preliminary pressure of 10 MPa was created. Then the pressure was increased to the required value, temperature being constant (250 K). When required pressure was reached and stabilized, the step-by-step cooling was realized, with temperature interval equal to 20 K.

Isobars of relative volume variation of ice sample as a function of temperature are shown in Figure 2 for the pressures 300, 500, 750, 1400 MPa. Phase transitions in the process of cooling are clearly seen on this isobars.

In Figures 3 and 4 there are shown $\Delta V/V_0$ versus P isotherms at temperatures 90 and 106 K for growing pressure and isobars at 1200, 1400, 1700, 2150 MPa for lowering temperature. On these curves transitions of Ih and VI ice modifications into ice XII modification (BIZHIGHITOV, SIROTA) are clearly seen.

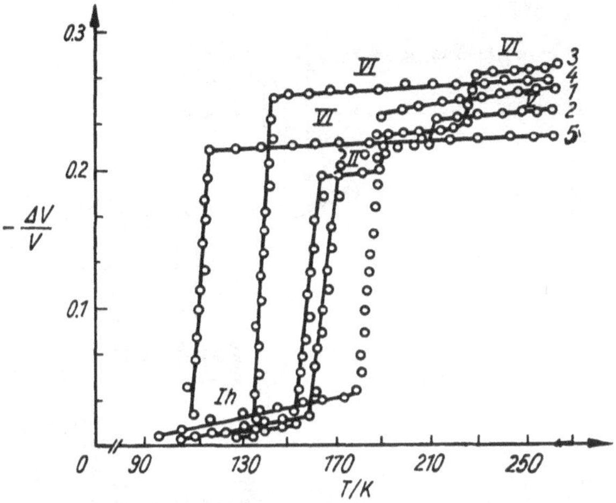

Fig. 2. Isobars of relative volume variation of ice sample for growing temperatures, for different pressures. Curves:
1: 300 MPa, 2: 500 MPa, 3: 750 MPa, 4: 1100 MPa, 5: 1400 MPa

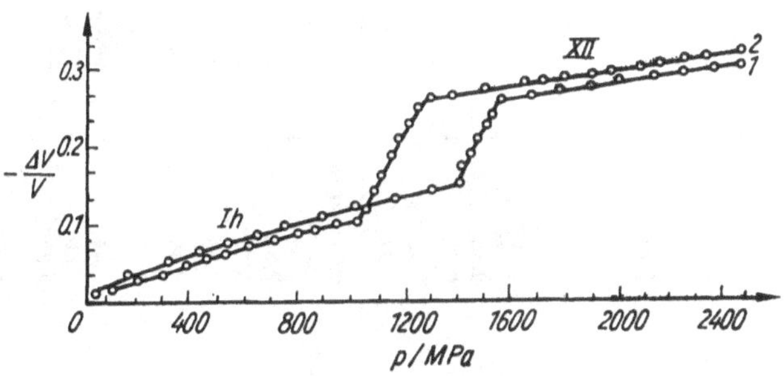

Fig. 3. Isotherms of relative volume variations for growing pressure at temperatures 103 K (1) and 90 K (2)

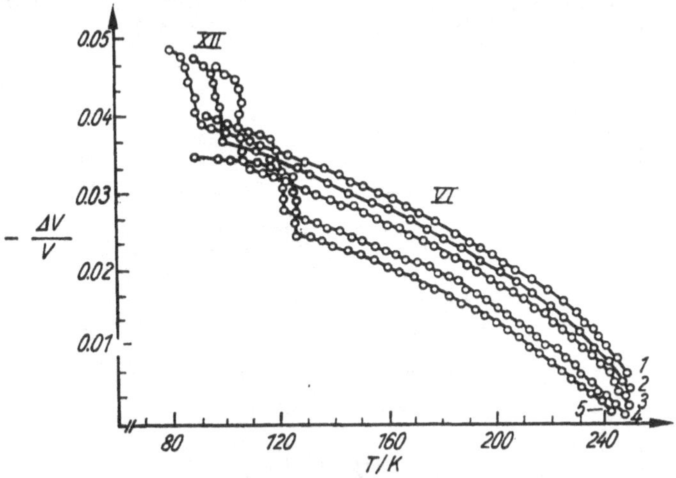

Fig. 4. Isobars of relative volume variations in process of cooling at pressures 1200 MPa (1), 1400 MPa (2), 1700 MPa (3), 2150 MPa (4), 2300 MPa (5)

Table 1
Isothermal measurements (T = const; P = const)

Transition	T_K K	P_K MPa	$\frac{dT_K}{dp}$, $\frac{K}{MPa}$	$\frac{V_0 - V}{V_0}$	$(V_0 - V) \times 10^{-6}$ m³/mole	ΔS J/mole	ΔH J/mole	$P\,\Delta V$ J/mole	ΔU J/mole	ΔG J/mole
1	2	3	4	5	6	7	8	9	10	11
					T = const					
Ih – III	253	200	−1.00	−0.160	−3.85	3.21	802.5	−770	1572.50	0
Ih – II	253	200	−0.80	−0.190	−3.92	4.90	1127	−784	1911	0
	213	290	−0.40	−0.220	−3.89	9.73	2037	−972.5	3009.50	0
II – V	233	420	−0.20	−0.020	−0.35	1.75	402.50	−147	549.50	0
	213	510	−0.10	−0.011	−0.27	2.70	567	−135	702	0
III – V	253	400	−0.40	−0.062	−1.13	2.83	707.50	−395.5	1103	0
V – VI	253	650	−0.20	−0.040	−0.90	4.50	1125	−558	1683	0
	233	800	−0.27	−0.031	−0.70	2.63	604.9	−560	1164.90	0
	213	830	−0.27	−0.032	−0.56	2.10	441	−453.6	894.60	0
Ih – XII	103	1450	0.05	−0.10	−2.10	−42.0	−4200	−3151	−1050	0
	90	1160	0.05	−0.13	−2.66	−53.20	−4788	−3192	−1596	0
VI – VII	300	2200	0.80	−0.04	−0.84	−0.05	−15.00	−1848	1883	0
	315	2250	0.80	−0.04	−0.91	−0.05	−15.50	−2047.50	2031.50	0
					P = const					
Ih – II	190	300	−0.18	−0.180	−3.85	21.29	4064.10	−1155	52.19	0
	168	500	−0.08	−0.170	−3.64	45.50	7644.00	−1820	9464	0
	158	750	−0.04	0.182	−3.86	55.14	8712.12	−2895	11607.12	0
II – V	213	500	−0.09	−0.020	−0.39	4.33	910	−195	1105	0
V – VI	233	750	−0.27	−0.030	−0.63	2.33	540.56	−472.50	1013.06	0
Ih – VI	142	1100	−0.13	−0.25	−5.26	40.46	5745.32	−5786	1153.30	0
	115	1400	−0.05	−0.28	−5.89	117.80	13429.20	−8246	21675.20	0
	128	1200	−0.08	−0.30	−5.90	73.75	9433.60	−7080	16513.60	0
VIII – XII	120	2150	0.06	−0.06	−1.20	−20	−2400	−2580	180	0
	130	2300	0.06	−0.06	−1.27	−21.17	−2730	−2921	191	0
VI – XII	108	1700	0.06	−0.01	−0.21	−3.50	−378	−357	−21	0
	100	1400	0.04	−0.01	−0.21	−5.25	−525	−294	−231	0
	90	1200	0.04	−0.01	−0.21	−5.25	−472.50	−252	−220.50	0

In the Table 1 there are given the temperature T_k, the relative specific volume change $\Delta V/V_0$ for observed phase transitions of ice in case of isothermal pressure variations and isobaric temperature variations.

Phase diagram, plotted according to the data given partly in Table 1, is presented in Figure 5.

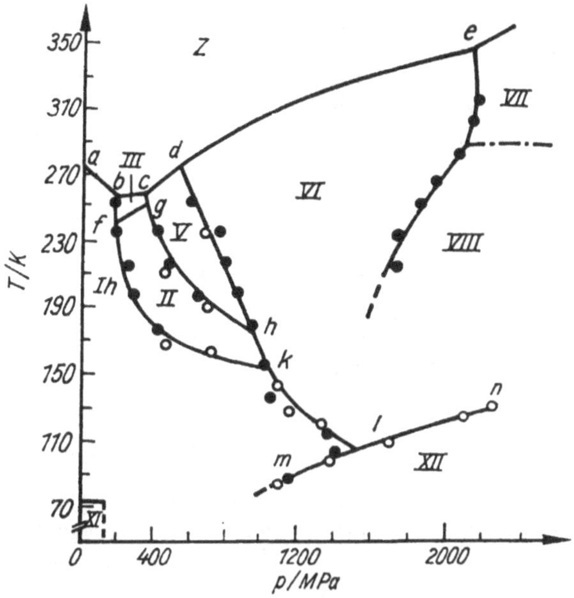

Fig. 5. Phase $P-T$ diagram of ice ● experimental points found according to relative volume variation in the process of isothermal growth of pressure; ○ experimental points found for the process of isobaric cooling

Equilibrium lines between water and ice (a b c d e) are plotted in accordance with the already known phase diagram (e.g. HOBBS). According to PISTORIUS et al. further increase of pressure leads to the growth of melting temperature of ice up to 715 K at 22000 MPa.

Equilibrium lines between different ice modifications with experimental points on them are plotted according to our data. Equilibrium line between VII and VIII modifications, shown on the phase diagram as a horizontal dot-and-dash line, has not been found in our experiment. It is drawn after the phase diagram by e.g. HOBBS.

On the phase diagram of Figure 5 line m—l—n denotes the transition of Ih and VI modifications into earlier unknown modification XII discovered in our experiment. The above transition line is plotted according to the measurements of specific volume variations in the process of isobaric cooling and in case of isothermal growth of pressure.

Using Clausius-Clapeyron equation

$$\frac{dT_k}{dP} = \frac{V_1 - V_2}{S_1 - S_2} = T_k \frac{V_1 - V_2}{H_1 - H_2},$$

we calculated variations of entropy $\Delta S = S_1 - S_2$, enthalpy $\Delta H = H_1 - H_2$, internal energy $\Delta U = U_1 - U_2 = \Delta H - P\,\Delta V$ per mole for the values of pressure and temperature of polymorphic transitions of ice modifications shown on the phase diagram of Figure 5.

According the parts of the diagram studied earlier, our results coinside comparatively well with the data of Bridgman on the variation of specific volume in phase transitions when $T > 230$ K and $P < 2500$ MPa. However, hypothetic form of phase boundaries, supposed earlier, differs from that established in our experiment, including the sign of dT_K/dP.

According to the results of present experiment for the equilibrium lines between modifications Ih—II, II—V, V—VI, I—VI dT_k/dP is negative ($dT_k/dP < 0$); for the equilibrium lines between modifications VI—VII (VI—VIII), Ih—XII, VI—XII Td_k/dP is positive. Values of ΔH, ΔS, ΔG for the named phase transitions, calculated for isobars and isotherms, are given in Table 1. On the phase diagram (Fig. 5) there are shown non-variant points f, g, h, k, l of three-phase equilibria, their coordinates being given in Table 2.

Table 2

Equilibrium phases	points	T/K	P/MPa
Ih—II—III	f	243	200
II—III—V	g	252	350
II—V—VI	h	173	970
Ih—II—VI	k	153	1070
Ih—VI—XII	l	105	1550

We estimated the compressibility and the modulus of dilatation of studied ice modifications as a function of pressure and temperature using the data on sample volume variation in case of isothermal growth of pressure. In Tables 3 and 4 the

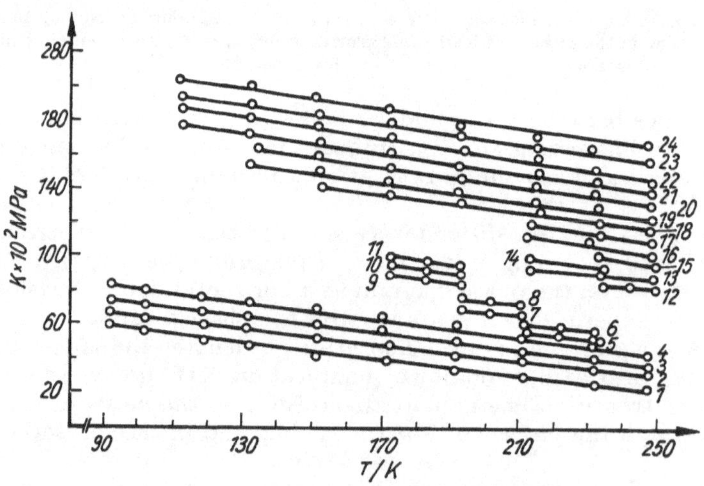

Fig. 6. Dependence of bulk moduli of ice modifications Ih, II, V, VI on temperature at different pressures:

ice Ih	50 MPa (1)	100 MPa (2)	150 MPa (3)	200 MPa (4)
ice II	300 MPa (5)	350 MPa (6)	400 MPa (7)	450 MPa (8)
	500 MPa (9)	600 MPa (10)	650 MPa (11)	
ice V	500 MPa (12)	550 MPa (13)	600 MPa (14)	700 MPa (15)
ice VI	950 MPa (16)	1050 MPa (17)	1150 MPa (18)	
	1250 MPa (19)	1350 MPa (20)	1450 MPa (21)	
	1550 MPa (22)	1650 MPa (23)	1750 MPa (24)	

Table 3

Phase	T K	P MPa	K × 10² MPa
Ih	248	15	20
		45	23
		73	26
		105	28
		135	36
		165	40
		195	43
	233	15	24
		45	26
		75	29
		105	34
		135	38
		165	43
		195	50
	213	15	27
		45	29
		75	34
		105	38
		135	42
		165	50
		195	56
	193	15	30
		45	33
		75	40
		105	45
		135	50
		175	53
		205	56
	173	15	33
		45	37
		75	43
		105	50
		135	56
		165	60
		195	65
		225	70
		255	75
		285	82
		305	88
		335	90
		315	93

Phase	T K	P MPa	K × 10² MPa
Ih	153	150	61
		250	81
		350	96
		450	111
		550	130
		650	147
		750	155
		850	162
		950	170
	133	150	65
		250	85
		350	103
		450	115
		550	133
		650	146
		750	158
		850	163
		950	170
	103	150	69
		250	88
		350	105
		450	120
		550	138
		650	151
		750	165
		850	169
		950	185
		1050	190
		1150	200
		1250	212
	93	150	71
		250	90
		350	107
		450	125
		550	140
		650	155
		750	170
		850	180
		950	186
		1050	190
		1150	200

Phase	T K	P MPa	K × 10² MPa
II	233	250	42
		280	45
		310	50
		340	54
		370	56
	213	300	54
		350	60
		400	65
		450	72
		500	77
	193	450	71
		500	83
		550	87
		600	89
		650	94
	173	500	83
		550	90
		600	95
		650	100
		700	105
		750	107
		800	111
		850	117
		900	123
		950	129
		1000	133
III	248	250	34
		260	31
		270	42
		280	45
		290	50
	248	415	75
		445	78
		475	79
		505	84
		535	86
		565	89
		595	92
	193	750	118
		800	124
		850	128

Phase	T K	P MPa	K × 10² MPa
V	233	500	83
		515	85
		545	90
		515	94
		605	98
		636	100
		695	105
		705	108
	213	550	94
		600	100
		650	106
		700	110
		750	115
	248	950	100
		1050	110
		1350	131
		1450	156
		1550	143
		1650	158
		1750	169
VI	233	950	106
		1050	115
		1150	121
		1250	128
		1350	136
		1450	145
		1550	153
	213	950	115
		1050	119
		1150	125
		1250	132
		1450	150
		1550	156
		1650	167
	193	1150	128
		1250	137
		1350	149
		1450	156
		1550	163
		1650	171

Phase	T K	P MPa	K × 10² MPa
VI	173	1150	130
		1250	144
		1450	158
		1550	168
		1650	175
		1750	186
	153	1150	140
		1250	158
		1350	161
		1450	166
		1550	178
		1650	184
		1750	194
	133	1250	132
		1350	164
		1450	170
		1550	180
		1650	188
		1750	196
		1850	208
	113	1450	176
		1550	184
		1650	192
		1750	202
	248	1900	222
		2000	227
		2100	232
		2200	250
VIII	233	1850	232
		1950	238
		2050	250
		2150	263
	103	1700	250
		1900	263
		2100	285
		2300	303
XII	90	1400	243
		1600	256
		1800	270
		2000	290
		2200	310

magnitudes of bulk modulus and its pressure and temperature derivatives for studied ice modifications for different temperatures and pressures are given.

Moduli of bulk for different modifications versus temperature and pressure are shown in Figures 6 and 7.

According to our data, bulk moduli for various modifications depend practically

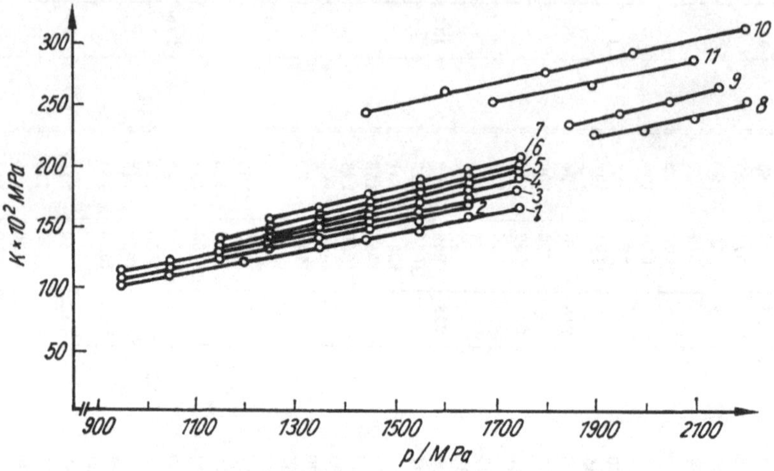

Fig. 7. Dependence of bulk moduli of ice modifications VI, VIII, XII on pressure at different temperatures:

ice VI	248 K (1)	233 K (2)	193 K (3)	
	173 K (4)	153 K (5)	133 K (6)	113 K (7)
ice VIII	248 K (8)	233 K (9)		
ice XII	90 K (10)	106 K (11)		

Table 4

Phase	P/MPa	T/K	φ	γ'	$\partial K/\partial P$	Phase	T/K	P/MPa	$\partial K/\partial T$
Ih	300	193	0.180	0.256	0.180	Ih	200	50	0.200
		173	0.170	0.251	0.170			100	0.201
		153	0.170	0.251	0.170			150	0.211
		133	0.160	0.246	0.160			200	0.213
		103	0.155	0.244	0.164	II	185	500	0.200
		93	0.150	0.241	0.150			600	0.260
II	450	213	0.100	0.216	0.093			650	0.300
		193	0.095	0.214	0.095	V	240	500	0.340
		173	0.094	0.213	0.100			550	0.360
V	550	248	0.100	0.216	0.100			600	0.380
		233	0.100	0.216	0.100			650	0.382
		213	0.100	0.216	0.100			700	0.400
		193	0.110	0.221	0.110	VI	155	1150	0.220
VI	1400	248	0.080	0.206	0.080			1250	0.231
		233	0.080	0.206	0.080			1350	0.264
		213	0.080	0.206	0.080			1450	0.280
		193	0.076	0.204	0.076			1550	0.300
		173	0.080	0.206	0.080			1650	0.322
		153	0.080	0.206	0.080			1750	0.331
		133	0.080	0.206	0.080	XII	100	1600	0.493
		113	0.080	0.206	0.080			1700	0.500
VIII	2050	248	0.075	0.204	0.075			1800	0.511
		233	0.074	0.203	0.074			1900	0.513
XII	1900	103	0.073	0.203	0.073			2000	0.516
		90	0.070	0.201	0.070			2100	0.518

linearly upon pressure and temperature, their pressure $\varphi_p = \left(\frac{\partial K}{\partial P}\right)_T$ and temperature $\varphi_T = \left(\frac{\partial K}{\partial T}\right)_p$ derivatives being practically constant for each phase.

Therefore, for phase j we may assume

$$\left(\frac{\partial \ln K}{\partial \ln V}\right)_{T,j} = -\varphi_{T,j}\,.$$

In Debye approximation characteristic Debye temperature $\theta_D = \frac{h\nu_m}{KT}$ is connected with bulk modulus: $\theta_D = \frac{h}{k}\left(\frac{9N}{4\pi}\right)^{1/3} \times \frac{K^{1/2}V^{1/6}}{M^{1/2}f(\mu)^{1/2}}$ (JAMES) where $f(\mu) = \left[\frac{1+\mu}{3(1-2\mu)}\right]^{3/2} - 2\frac{2(1+\mu)^{3/2}}{3(1-2\mu)}$ and Poison coefficient μ virtually does not depend on the volume.

Supposing $\frac{1}{2}\frac{\partial \ln f(\mu)}{\partial \ln V}$ to be constant, we have:

$$\gamma_j' = -\left(\frac{\partial \ln \theta}{\partial \ln V}\right)_j = -\left(\frac{1}{2}\frac{\partial \ln K}{\partial \ln V} + \frac{1}{6}\right) = \frac{1}{2}\varphi_{T,j} - \frac{1}{6},$$

where γ_j' is analogous to Grüneisen constant. Its values for some modifications are given in Table 4.

4. Conclusions

Study of specific volume variation of ice at low temperatures and comparatively high pressures allowed to make substantial alterations in the earlier supposed hypothetic phase diagram of ice and to construct it anew in the range of temperature lower than 250 K at pressures up to 2500 MPa. A new modification — ice XII — was found in our experiment existing in significant area of phase diagram under the line m—l—n, between the points 90 K at 1200 MPa and 120 K at 2150 MPa.

Our study of the influence of pressure and temperature upon the variation of specific volume at temperatures from 250 to 290 K and pressures up to 2500 MPa allowed to evaluate the bulk moduli for different modifications. It was shown that bulk modulus increases as pressure grows, as temperature falls, and if we proceed to denser modifications.

References

BIZHIGHITOV, T. B., SIROTA, N. N.: Zh. Eksp. Teor. Fiz. (pisma) **44** (1986) 324

BRIDGMAN, P. W.: a) Proc. Amer. Acad. Arts. Sci.: **47** (1912) 441; b) J. Chem. Phys. **3** (1935) 597; c) J. Chem. Phys. **5** (1937) 964

HOBBS, P. V.: Ice Physics, London, Oxford 1974

JAMES, R. W.: The optical principles of the diffraction of X-Rays, London 1950

LIN-GUN-LIU: Earth and Planetary Sci. Lett. **61** (1982) 359

PISTORIUS, C. W., PISTORIUS, M. C., BLACKY, J. D., ADMIRAL, L. J.: J. Chem. Phys. 38 (1963) 600
POLIAN, A., GRIMSDITCH, M.: Phys. Rev. 52 (1984) 1312
SUGA HIROSHI: Solid State Phys. 20 (1985) 125
TAMMANN, G.: Ann. Phys. 2 (1900) 1

(Received, accepted October 27, 1987)

Authors' address:

Prof. Dr. N. N. SIROTA, T. B. BIZHIGHITOV
MGMI, Department of Physics
ul. Pryanishnikova, 19
125008 Moscow, USSR

Cryst. Res. Technol.	23	1988	5	605–608

A. KRAFT, B. GREULING

Akademie der Wissenschaften der DDR, Zentralinstitut für Physik der Erde, Potsdam

High Pressure Phase Transformation in MnS

The structural, optical and electrical behaviour of MnS under high pressure was investigated. For X-ray investigations and optical observations a gasketed diamond anvil cell was used. Measurements of the electrical conductivity were performed in a Bridgman opposed-anvil apparatus. A structural phase transition at 7.2 GPa from the cubic α-phase (NaCl-type) to the B16-type structure (GeS-type) was detected by X-ray investigations and confirmed by optical observations. Measurements of the electrical conductivity showed a linear decreasing electrical gap energy in dependence on pressure.

Es wurde das strukturelle, optische und elektrische Verhalten von MnS unter Druck untersucht. Für die Röntgenuntersuchungen und die optische Beobachtung wurde eine gasketierte Diamantstempelkammer verwendet. Die Messung der elektrischen Leitfähigkeit wurde in einer Bridgman-Flachstempelapparatur durchgeführt. Ein struktureller Phasenübergang bei 7.2 GPa von der kubischen α-Phase (NaCl-Typ) in die B16-Struktur (GeS-Typ) wurde durch Röntgenuntersuchungen festgestellt und durch optische Beobachtungen bestätigt. Die Messung der elektrischen Leitfähigkeit zeigte eine lineare Abnahme der elektrischen Gap-Energie in Abhängigkeit vom Druck.

Introduction

The phase relations in the Earth's lower mantle are subjects of considerable interest and controversy. The major phases below about 650 km have been proposed to be (Mg, Fe)SiO_3 with perovskite structure and (Mg, Fe)O with B1-type structure (NaCl-type). Investigations of oxides and sulphides with MX-stoichiometry have resulted in phase transformations different to the expected B1—B2 transition. Interest has recently arisen by a paper of NAVROTSKY and DAVIES. They suggested that the B8-type structure (NiAs-type) is another possible candidate for (Mg, Fe)O at very high pressure and calculated the transition pressures for several substances with B1-type structure which are supposed to favour the NiAs-type structure. The lowest pressures resulted for the manganese chalcogenides. CLENDENEN and DRICKAMER observed a volume change for MnS near 10 GPa and calculated a distorted NaCl-type structure. On the basis of their results for MnSe ČEMIČ and NEUHAUS assumed that transition in MnS to be of the type B1—B8. We therefore took MnS as a model substance in order to investigate this transition.

Experimental procedure

Samples of α-MnS were prepared by the evacuated silica glass tube technique at temperatures of 800 °C. Powdered samples were investigated in a diamond anvil cell up to 20 GPa by high pressure X-ray diffraction. The DAC was equipped with diamond anvils of 0.5 mm anvil flat diameter and loaded with a gasket made of EP 570 steel. It was then prepressed and a hole of 0.15 mm diameter was drilled into the gasket. The hole was filled with MnS powder, with ruby as a pressure sensor and a methanol ethanol mixture as a pressure transmitting medium. A gasketed diamond anvil cell has also been used for optical observations. The pressure was measured by the well-known ruby fluorescence technique.

The electrical conductivity measurements were carried out in a Bridgman opposed-anvil apparatus with an external heater which is described elsewhere (Seipold, Witzki). The anvils were made of WC/Co which permits pressures up to 8 GPa and temperatures from room temperature up to 300 °C. The pressure was calibrated against the polymorphic phase transitions of bismuth at 2.55 and 7.7 GPa and of barium at 5.53 GPa. The temperature was measured by using a Ni/CrNi-thermocouple which is fixed at the anvils.

Structural transition

The first signal of a phase transition was a colour change during an optical observation under the microscope. In the region between 8 and 10 GPa the colour of the sample changed from initial yellow to red and with increasing pressure the red colour became darker up to black.

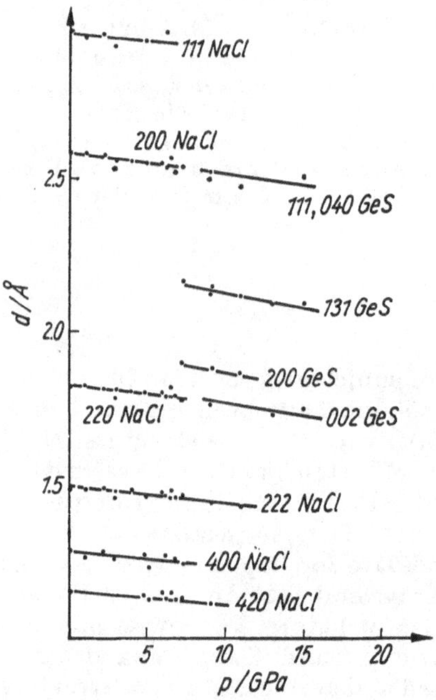

Fig. 1. Interplanar distances vs pressure

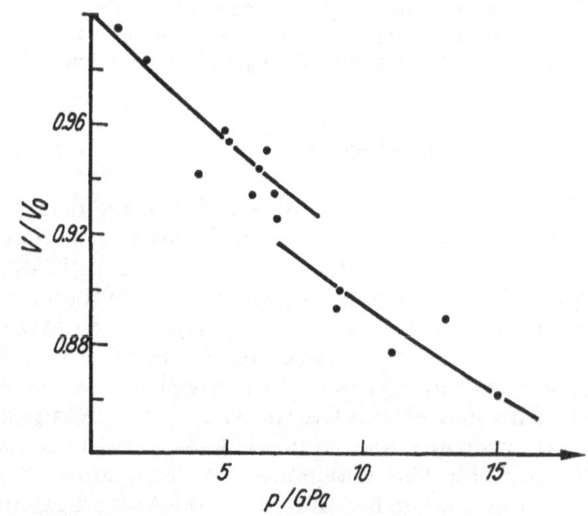

Fig. 2. Pressure dependence of the unit cell volume up to 15 GPa (at the top of the ordinate read 1.00)

In the X-ray diffraction pattern new peaks appear at 7.2 GPa. The variation of the interplanar distances of MnS as a function of pressure up to 15 GPa is shown in Figure 1. We tried to index this new pattern on the basis of cubic, tetragonal, hexagonal, and orthorhombic symmetry and to compare it with related structure. The best results could be obtained by assuming a transition to the B16-type structure (GeS-type). This transition was also found in lead chalcogenides (Chattopadyay et al.). The B1- and the B16-type structure are closely related. They are explained geometrically by a translation of $a/2$. The transition in MnS exhibits a relative volume change of 1.6% (Fig. 2).

Electrical measurements

The pressure dependence of the electrical conductivity of MnS showed a nearly exponential decrease with increasing pressure. This is the behaviour of many semiconductors before the phase transition. Also the temperature dependence of the conductivity at constant pressure shows semiconducting behaviour. In Figure 3 a log R vs $1/T$ plot for α-MnS exhibits the expected linearity. From

$$R(T) = R_0 \exp (E_A/2kT) \quad (1)$$

one can estimate the activation energy E_A. Considering the linear decreasing activation energy with increasing pressure which is shown in Figure 4 one can suppose that at about 10 GPa the gap will be closed. That value corresponds to the optically observed semiconductor-metal transition.

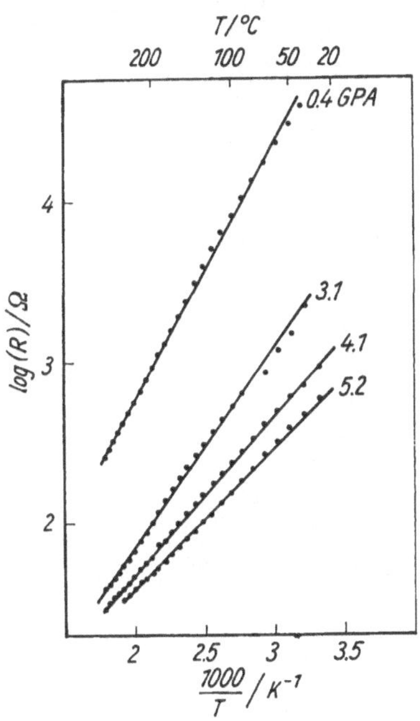

Fig. 3. Temperature dependence of the resistance at various pressures (read GPa)

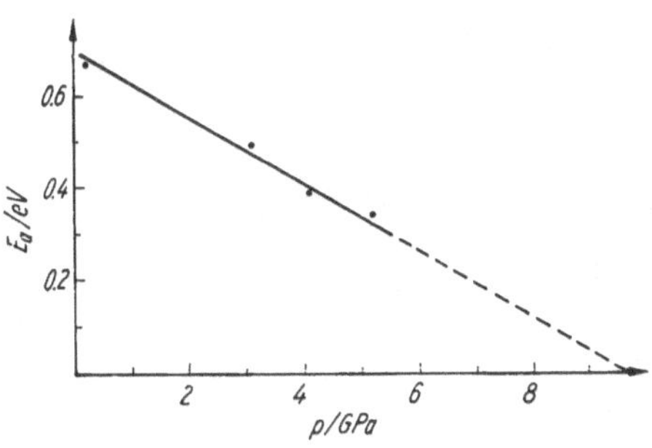

Fig. 4. Activation energy vs pressure

Conclusions

The expected phase transition from the B1-type to the B8-type structure could not be confirmed. In MnS a phase transition to the B16-type structure takes place at 7.2 GPa which is connected with a volume change of 1.6% and a colour change from yellow to red up to black. MnS shows semiconducting behaviour up to 8 GPa. We suppose a semiconductor-metal transition because of the linear decreasing gap energy and the optically observed colour change. We assume a high pressure structural behaviour similar to the lead chalcogenides.

References

CEMIC, L., NEUHAUS, A.: High Temp.-High Press. **6** (1974) 203
CHATTOPADHYAY, T., WERNER, A., VON SCHNERING, H. G., PANNETIER, J.: Rev. Phys. Appl. **19** (1984) 807
CLENDENEN, R. L., DRICKAMER, H. G.: J. Chem. Phys. **44** (1966) 4223
NAVROTSKY, A., DAVIES, P. K.: J. Geophys. Res. **86** B5 (1981) 3689
SEIPOLD, U., WITZKI, H.: Exp. Technik Phys. **28** (1980) 261

(Received, accepted November 9, 1987)

Authors' address:

Dr. A. KRAFT, B. GREULING
Zentralinstitut für Physik der Erde
Telegrafenberg
Potsdam
DDR-1561

Cryst. Res. Technol.	23	1988	5	609–619

Yu. D. Chistyakov, Yu. A. Baikov, M. V. Akulionok

Moscow Institute of Electronic Engineering, Moscow, USSR

Fluctuation Theory of the One-component and Binary Metallic Melt's Crystallization in the Case of Micro- and Macro-Systems

All types of fluctuation mechanisms resulting in a normal kinetics of the one-component and binary melt's crystallization within a small supercooling region for micro- and macro-systems are systematized. Three principal mechanisms of spontaneous fluctuations acting in the monolayers of two-phase transitional zone in the vicinity of melting temperature for one-component metallic crystals and of the liquidus temperature for binary ones are described. The formulas of kinetic coefficients figuring in the normal law of crystallization conformably to one-component and binary metallic micro- and macrosystems for all types of fluctuation mechanisms are presented. Some peculiarities of each fluctuation mechanism are presented. Some peculiarities of each fluctuation mechanism acting within a small supercooling region of the melt-crystal micro- and macrosystems are pointed out. A comparison of theoretical and experimental data concerning the one-component and binary melt's crystallization kinetics at some small supercoolings is presented.

В данной работе систематизированы все разработанные виды флуктуационных механизмов, ведущие к нормальной кинетике процессов кристаллизации металлических однокомпонентных и бинарных расплавов в области малых переохлаждений для микро- и макросистем. В статье приведены три основные механизма спонтанных флуктуаций концентраций частиц твёрдого состояния, действующие в монослоях переходной двухфазной зоны в области малых отклонений от температур плавления для однокомпонентных металлических кристаллов и от температур линии ликвидус для бинарных систем. Приведены явные выражения для величин кинетических коэффициентов, фигурирующих в законе нормального роста кристаллов, для всех видов флуктуационных механизмов применительно для однокомпонентных и бинарных металлических микро- и макросистем. Отмечены особенности каждого из механизмов в случаях их действия в области малых переохлаждений системы расплав-кристалл на микро- и макроскопическом уровнях. Проведено сравнение теоретических и экспериментальных данных по кинетике кристаллизации однокомпонентных и бинарных макросистем в области малых переохлаждений системы расплав-кристалл.

Introduction

The one-component and binary metallic melt's crystallization within a small supercooling region (not more than 1°) are connected first of all with crystal growth kinetics. Very high crystallization rates at such metals as Ga, Hg, Cd, In, K, Rb, Cs, Sn, Pb and also at refractory ones Fe, Co, Ni in the vicinity of their melting temperatures T_0 can be explained only by the so-called normal crystallization law, when the average rate of metallic solid phase formation V is in direct ratio to first degree of the melt-crystal system's supercooling ΔT ($V \cong K\,\Delta T$, where K is a kinetic coefficient). In the case of above-mentioned metal crystallization and also of binary metallic systems, for example, In + 9 at.% Bi, In + 4.6 at.%Pb, In + 7.1 at.% Pb, In + 9.5 at.% Pb, Ga + 8 at.% Sn and others which have numerous experimental data on kinetics of crystal growth (see BORISOV et al. 1965, 1969; BORISOV, MATVEEV) a presence of two-phase transitional zone and of so-called spontaneous thermal equilibrium-like fluctuations of the solid state particle concentrations in each monolayer of given transitional area are specific and confirmed by available theoretical and experimental evidence (BORISOV; GILMER, SWALIN). For the last years a new applied analytic theory of normal crystal growth in the case of one-component and binary melt's crystallization within a small supercooling region has been developed. The theory is based on different types of spontaneous fluctuations of the solid state particle concentrations. All the below described fluctuation mechanisms acting within a small supercooling region of the metallic melt-crystal system lead to normal crystal growth kinetics. At first the fluctuation theory of normal crystal growth has been applied for the metallic one-component and binary micro-systems crystallizing from proper melts when so-called macroscopic parameter N (a full number of particles of both liquid and solid states in each monolayer of two-phase transitional zone) is not more than 10^3. Subsequently the given theory has been developed for the one-component and binary macro-systems (parameter tends to infinity $N \to \infty$). The purpose of the present paper is a comparison of earlier worked out mechanisms of spontaneous fluctuations for all the cases of one-component and binary micro- and macro-systems, moreover, it is necessary to point out the general properties and specific peculiarities of each mechanism considered acting within a small supercooling region.

1. Mechanism of spontaneous fluctuations of the "unlimited" spectrum solid state particle concentrations in the cases of one-component micro- and macro-systems

A model of given mechanism in a scheme of a two-phase transitional zone presented in concentration symbols is shown in Figure 1. Principal peculiarity of this mechanism is that as a result of spontaneous fluctuation which is some arbitrary move of the

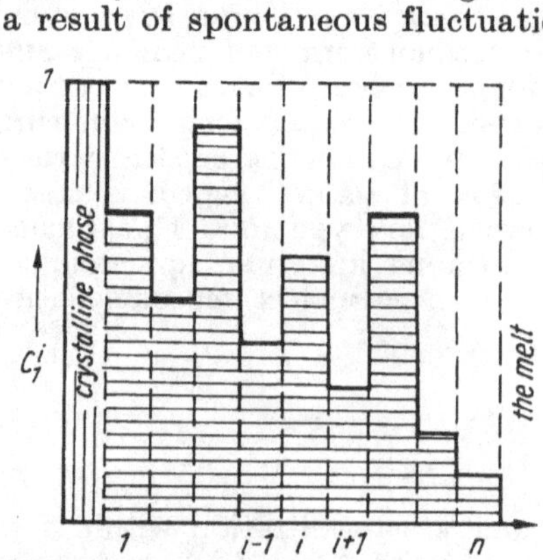

Fig. 1. A physical scheme in the concentration symbols of two-phase transitional zone consisting of n monoatomic thickness' layers; to the left of it is a homogeneous crystalline phase, to the right the melt. The hatched columns in each i-monolayer ($i = 1, \dots, n$) are the solid state particle concentrations $C_i = m_i/N$ (m_i – a number of particles being in a crystalline phase), the unhatched ones are the liquid state particle concentrations. The given scheme presents a model of spontaneous fluctuations of the "unlimited" spectrum solid state particle concentrations

upper level of solid state particle concentration from an initial (pre-fluctuation) position into a final (post-fluctuation) one, the solid state particle concentration in each monolayer of the above mentioned transitional zone can be of any physically possible value from zero (the pure melt) to unit (pure crystalline phase). The given move is in each monolayer of two-phase zone for an average time "τ" called the fluctuation period. The values are constant for each monolayer in question and generally speaking depend on the crystallizing one-component or binary metallic system's temperature. At first this mechanism has been developed for one-component metallic microsystems crystallizing from the proper melts. As shown by BAIKOV et al. (1978a, 1979a, 1981), the given mechanism of spontaneous fluctuations of the "unlimited" spectrum solid state particle concentrations leads to normal kinetics of crystal growth: $V \cong K\,\Delta T$, where K — kinetic coefficient — can be written as follows: $K \cong \frac{a}{\tau}\frac{q}{kT\,T_0}\,\Phi(N, \theta)$

$$\Phi(N, \theta) = \sqrt{\frac{|A|}{\pi}}\left[1 - \exp\left(-\frac{|A|}{4}\right)\right]\Big/|r + \theta|,\; A = N(\theta + r)\,. \tag{1.1}$$

Here: a is a mean interatomic distance for metallic system considered, q the melting heat per atom, T a current temperature of the melt-crystal system, $\theta = \frac{2}{3}\frac{q}{kT}$ the so-called roughness parameter, k the Boltzmann constant, $r = -4 \ln 2$, $\Delta T = T_0 - T$.

Subsequently the above-stated mechanism of spontaneous fluctuations has been approximated to one-component metallic macrosystems (N — parameter tends to ∞) as a result a compact formula describing the normal crystallization law of one-component metals in the vicinity of melting temperatures is $V^* = \lim_{N\to\infty} V = K^*_{\text{eff}}\,\Delta T$ where the effective kinetic coefficient K^*_{eff} is

$$K^*_{\text{eff}} \cong \frac{a}{\tau}\,\frac{3q}{(8 \ln 2)\,kT\,T_0}\,. \tag{1.2}$$

The given formula permits to estimate the average crystallization rates of the one-component metallic melts within a small supercooling region and to compare them with available experimental data for some particular metals, for example, Ga, Hg, Cd, K, Rb, In (see Tab. 1). As seen from the Table 1 the best agreement between the theoretical and experimental values exists for Ga, Hg, Cd, In. For the alkaline metals K and Rb which have extensive transitional zones a coincidence of theory and experiment is less successful. The τ-values for the given metals have been obtained by virtue of the works of LEAK and SWALIN, and SWALIN 1959, 1961. The values for Hg, Cd, K, In and for Ga, Rb are $\tau = 10^{-11}$ s, $\tau = 10^{-10}$ s, respectively. When transiting to macroscopic systems i.e. at. $N \to \infty$, the fluctuation mechanism in question in each monolayer of two-phase zone deteriorates into only one possible spontaneous fluctuation as a result of which the solid state particle concentration becomes equal to about $\frac{1}{2}$ (see BAIKOV, CHISTYAKOV 1985a). As seen from the formulas (1.1), (1.2) when acting the fluctuation mechanism of the "unlimited" spectrum solid state particle concentrations the average crystallization rates of the one-component metallic melts within the small supercoolings' region do not depend on two-phase transitional zone's size (parameter "n") for both micro- and macrosystems. This is a peculiarity of given mechanism. The kinetic coefficient's dependence on the macroscopic parameter N is only at the metallic microsystems.

Table 1
A comparison of theoretical and experimental data concerning the melts' crystallization kinetics for Ga, Hg, Cd, K, Rb, In-metals within the small supercooling region. The relative errors of experimental data $(\Delta V_{exp}/V_{exp}) \times 100\%$ for the given metals are: 1.9%; 1.1%; 2.0%; 2.4%; 2.0%; 9.1% respectively

Metal	ΔT/K	V^*/mm · s^{-1}	V_{exp}/mm · s^{-1}
	0.1	3.0	5.3
Ga	0.3	9.1	15.9
	0.5	15.2	26.5
	0.05	4.2	4.6
Hg	0.1	8.4	9.2
	0.5	42.0	46.0
	0.02	0.7	1.1
Cd	0.1	3.4	5.5
	0.3	10.3	16.5
	0.5	17.1	27.5
	0.01	0.7	4.0
K	0.05	3.6	20.0
	0.08	5.8	33.0
	0.26	2.1	7.5
Rb	0.28	2.3	9.6
	0.36	2.9	14.0
	0.05	1.9	5.5
In	0.1	3.7	11.0
	0.2	7.5	22.0

2. Mechanism of the ordered "shortened" spontaneous fluctuations of the solid state particle concentrations for one-component metallic micro- and macrosystems

The previous mechanism of spontaneous fluctuations does not establish unfortunately a correct shape of the two-phase transitional zone which as numerous experimental evidence support is step-like, i.e. as the number of monolayer increases, the mean value of solid state particle concentration gradually diminishes. As for results of above stated mechanism the values in all the monolayers of a two-phase transitional zone are equal about to $\frac{1}{2}$. To improve this drawback a mechanism of the ordered "shortened" spontaneous fluctuations conformably to one-component metallic microsystems (see BAIKOV et al. 1978b) and proper macrosystems crystallizing from the melts in the vicinity of the melting temperature has been worked out. According to this mechanism all the spontaneous fluctuations are realized in a strong order: the first fluctuation is in the $i = 1$ monolayer at $t_0 + \tau$ moment of discrete time, the second is in the $i = 2$ one at $t_0 + 2\tau$ moment of discrete time etc., the n-fluctuation is in the $i = n$ monolayer at $t_n = t_0 + n\tau$ moment of time (t_0 is an initial point of fluctuation). The processes of similar ordered spontaneous fluctuations in all the monolayers of the transitional zone are described by the same system of integral equations as well as for spontaneous fluctuations of the "unlimited" spectrum solid state particle concentrations (see BAIKOV et al. 1978a, 1979a, 1981), however, a basic three-dimensional conditional differential distribution function $W_3(X_{ii+1} \mid X_{i-1i}\, X_{i+1i})$

can be formulated in some "shortened" form as follows:

$$W_3(X_{ii+1} \mid X_{i-1i}X_{i+1i}) = \{\exp[-n_i(\eta_i - b_i)^2] / \int_0^1 [-n_i(\eta - b_i)^2]\, d\eta\} \times \\ \times \delta_{|m_i|\, m_i}\delta_{|p_i|\, p_i}/(X_{i-1i} - X_{i+1i}) \tag{2.1}$$

where:

$$\eta_i = (X_{ii+1} - X_{i+1i})/(X_{i-1i} - X_{i+1i}),\; n_i = N\overline{X}_{i-1i}\,|r + \theta\overline{X}_{i-1i}|\,,$$

$$b_i = \frac{1}{2}\left[1 - \frac{\delta + \theta(\overline{X}_{i-1}\,i - 1)}{r + \theta\overline{X}_{i-1i}}\right],$$

$$\delta = \frac{q}{kT}\frac{\Delta T}{T_0},\qquad \delta_{|m_i|m_i} = \begin{cases}1 & \text{at} \quad m_i > 0\\ 0 & \text{at} \quad m_i \leqq 0,\end{cases}\qquad \delta_{|p_i|p_i} = \begin{cases}1 & \text{at} \quad p_i > 0\\ 0 & \text{at} \quad p_i \leqq 0\end{cases}$$

$$m_i = X_{ii+1} - X_{i+1i},\; p_i = X_{i-1i} - X_{ii+1}.$$

Here: $X_{ij} = C_i(t_j)$, $\overline{X}_{ij} = \overline{C}_i(t_j)$ are the solid state particle concentration in an i-monolayer at $t_j = t_0 + j\tau$ moment of discrete time and the average value of the same concentration, respectively. By virtue of the (2.1) distribution function one can establish a correct step-like shape of transitional zone which should be defined by the $\{\overline{X}_{ii+1}\}$-set (see Fig. 2), where $\overline{X}_{ii+1} \cong (\frac{1}{2})^i$ $(i = 1, \ldots, n)$ within the small supercooling region and for a well developed transitional zone, when the condition $|\theta/r| \ll 1$

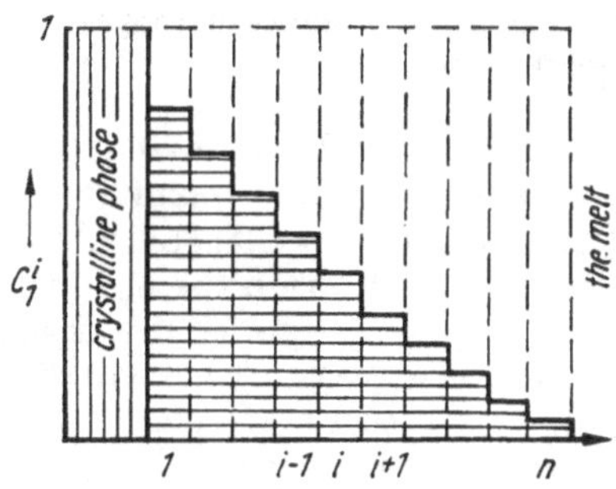

Fig. 2. A physical scheme in the concentration symbols of two-phase transitional zone consisting of n monoatomic thickness' layers; to the left there is a homogeneous crystalline phase, to the right the melt. The hatched columns in each i-monolayer ($i = 1, \ldots, n$) are the solid state particle concentrations $C_i = m_i/N$ (m_i – a number of particles being in a crystalline phase), the unhatched ones are the liquid state particle concentrations. The given scheme presents the model of spontaneous ordered "shortened" fluctuations and of spontaneous ones of the limited spectrum solid state particle concentrations

is just (see (Baikov et al. 1978a, b, 1979a, 1981). In the case of one-component metallic microsystems in the vicinity of melting temperatures given mechanism of the ordered "shortened" spontaneous fluctuation leads to the normal crystallization law:

$$V \cong \frac{a}{\tau}\frac{q}{kT\,T_0}\Phi(n, N, \theta)\,\Delta T\,, \\ \Phi(n, N, \theta) = \frac{1}{n\sqrt{\pi}}\sum_{i=1}^{n}\sqrt{n_i}\left[1 - \exp\left(-\frac{n_i}{4}\right)\right]\left|r + \theta\left(\frac{1}{2}\right)^{i-1}\right|^{-1}. \tag{2.2}$$

In the case of one-component metallic macrosystems crystallizing from the proper melt within a small supercooling region given mechanism results in the same normal kinetics with the following effective kinetic coefficient: $V^* = \lim_{N\to\infty} V \cong K^*_{\text{eff}}\, \Delta T$

$$K^*_{\text{eff}} = \frac{a}{\tau}\left(\frac{n-1}{8\ln 2}\right)\frac{q}{kT\,T_0}. \tag{2.3}$$

As seen from the (2.2) and (2.3) formulas the ordered "shortened" spontaneous fluctuation mechanism in the case of one-component metallic micro- and macro-systems' conditions the proper kinetic coefficients' dependence on two-phase transitional zone's size (the "n"-parameter). In the case of macrosystems ($N \to \infty$) all the fluctuation mechanism deteriorates into only one possible spontaneous fluctuation in each monolayer of the transitional zone. As a result the solid state particle concentration in the i-monolayer becomes equal to $\overline{X}_{ii+1} \cong (\frac{1}{2})^i$ ($i = 1, \dots, n$). For a given type of the fluctuation mechanism — as well as in the previous one — the effective kinetic coefficient K^*_{eff} does not depend on the macroscopic parameter N. Contrary to the fluctuation mechanism of the "unlimited" spectrum solid state particle concentration in the given case is a connection between the "n" and "N" parameter, namely: $n \ll 1 + 1.4 \ln N$. This presents the specific condition of above mentioned fluctuation mechanism's realization in the vicinity of the melt-crystal system's equilibrium temperature.

3. Mechanism of spontaneous fluctuations of the limited spectrum solid state particle concentrations when crystallizing one-component metallic melts

The given mechanism of spontaneous fluctuations is characterized by the following principle. As a result of one fluctuation, for example, in an i-monolayer ($i = 1, \dots, n$) at $t_k = t_0 + k\tau$ moment of discrete time (t_0 is a starting-point) the solid state particle concentration is less or more than the same concentrations in the ($i-1$)- and ($i+1$)-monolayers at pre-fluctuation point of time $t_{k-1} = t_0 + (k-1)\,\tau$, respectively, i.e. the following conditions are just: $X_{i+1k-1} \lesseqgtr X_{ik} \lesseqgtr X_{i-1k-1}$, where $X_{ik} = C_i(t_k)$, $X_{i+1k-1} = C_{i+1}(t_{k-1})$, $X_{i-1k-1} = C_{i-1}(t_{k-1})$ — the solid state particle concentrations in the i, ($i+1$), ($i-1$)-monolayers at t_k and t_{k-1} moments of discrete time, respectively. A model of similar spontaneous fluctuations for the case of one-component metallic microsystems crystallizing from melts has been stated in detail by BAIKOV et al. (1979a). The mechanism's application is valid if the above mentioned condition $n \ll 1 + 1.4 \ln N$ is fulfilled. The given mechanism of spontaneous fluctuations of the limited spectrum solid state particle concentrations is characterized by the step-like transitional zone shown in Figure 2. The transitional two-phase zone's shape is defined by the $\{\overline{X}_{ii}\}_{(i=k)}$ set where within the small supercooling region and for well developed transitional areas ($|\theta/r| \ll 1$) the average solid state particle concentrations are $\overline{X}_{ii} \cong (\frac{1}{2})^i$ ($i=1, \dots, n$). Within the small supercooling region ($(\Delta T/T_0) \ll 1$) the given mechanism as well as the two previous ones leads to the normal crystallization law which for the one-component metallic microsystems can be written as follows:

$$\begin{aligned} V &= \frac{a}{\tau}\frac{q}{kT\,T_0}\,\Phi(n, N, \theta)\,\Delta T \\ \Phi(n, N, \theta) &= \frac{1}{n\sqrt{\pi}}\sum_{i=1}^{n}\sqrt{n_i}\left[1 - \exp\left(-\frac{n_i}{4}\right)\right]\Big/\left|r + \theta\left(\frac{1}{2}\right)^{i-1}\right|. \end{aligned} \tag{3.1}$$

When coming to a one-component metallic macrosystems ($N \to \infty$) at $\Delta T/T_0 \ll 1$ the fluctuation mechanism of the limited spectrum solid state particle concentrations gives the same crystallization law with the following effective kinetic coefficient: $V^* = \lim_{N \to \infty} V \cong K^*_{\text{eff}} \Delta T$

$$K^*_{\text{eff}} \cong \frac{a}{\tau} \left(\frac{n-1}{8 \ln 2}\right) \frac{q}{kT\, T_0}. \tag{3.2}$$

As with two previous fluctuation mechanisms given one being applied to the macrosystems deteriorates into only one spontaneous fluctuation in each monolayer of the transitional zone. As a result the solid state particle concentration at $|\theta/r| \ll 1$ becomes equal to $\overline{X}_{ii} \cong (\frac{1}{2})^i$ $(i = 1, \dots, n)$. From the (3.2) formula a next conclusion follows: in the case of macrosystems the crystallization rate depends on two-phase transitional zone's size. As seen from the (2.2), (3.1), and (2.3), (3.2) formulas the mechanisms of

Table 2
A comparison of theoretical and experimental data concerning the melts' crystallization kinetics for Ga-, Hg-, Cd-, In metal within a small supercooling region. The relative errors of experimental data $(\Delta V_{\text{exp}}/V_{\text{exp}})$ 100% for the given metals are: 1.9%; 1.1%; 1.8%; 9.1%, respectively. Two-phase transitional zone's size is characterized by $n = 7$

Metal	ΔT/K	V^*/mm · s^{-1}	V_{exp}/mm · s^{-1}
	0.1	5.8	5.3
Ga	0.3	17.0	15.9
	0.5	29.0	26.5
	0.05	7.9	4.0
Hg	0.1	16.0	9.2
	0.5	79.0	46.0
	0.02	1.3	1.1
Cd	0.1	6.5	5.5
	0.3	19.0	16.5
	0.5	32.0	27.5
	0.05	3.7	5.5
In	0.1	7.2	11.0
	0.2	14.0	22.0

Table 3
A comparison of theoretical and experimental data concerning the melts' crystallization kinetics for K- and Rb metal within a small supercooling region. The relative errors of experimental data $(\Delta V_{\text{exp}}/V_{\text{exp}})$ 100% for given metals are: 2.4%; 2.0%, respectively. Two-phase transitional zone's sizes are characterized by $n = 30$

Metal	ΔT/K	V^*/mm · s^{-1}	V_{exp}/mm · s^{-1}
	0.01	6.8	4.1
K	0.05	34.0	20.0
	0.08	54.0	33.0
	0.26	20.0	7.5
Rb	0.28	22.0	9.6
	0.36	27.0	14.8

the ordered "shortened" spontaneous fluctuations and of spontaneous ones of the limited spectrum solid state particle concentrations within the small supercooling region and for well-developed transitional zones describe in the same manner the crystallization kinetics of both micro- and macrosystems. In the Tables 2 and 3 the average crystallization rates of some metals within the small supercooling region and a comparison with experimental evidence are presented for different sizes of the transitional zone. From the comparison of the Tables' 1, 2, 3 data one can see that transitional zone's size calculation furthers the better estimation of one-component metals' kinetics at least from the experimental point of view.

Thus, when acting all three fluctuation mechanisms considered in the case of one-component metallic microsystems the kinetic coefficients depend on the macroscopic parameter N. For the mechanisms of the ordered "shortened" spontaneous fluctuations and of spontaneous ones of the limited spectrum solid state particle concentrations the condition $n \ll 1 + 1.4 \ln N$ is desirable. A general point at all three fluctuation mechanisms for one-component metallic macrosystems is a degeneration of the fluctuation spectrum in question into only one spontaneous fluctuation in each monolayer of the transitional zone. All the effective kinetic coefficients don't depend on N-parameter. In the case of spontaneous fluctuations of the "unlimited" spectrum the kinetic coefficients don't depend on the n-parameter for both micro- and macrosystems. The n-parameter has an influence on the crystallization kinetics only in the cases of ordered "shortened" fluctuations and of the limited spectrum solid state particle concentration ones.

4. Mechanism of spontaneous fluctuations of the limited spectrum solid state particle concentrations when studying the metallic binary melts crystallization

The main ideas concerning the fluctuation mechanism of the limited spectrum solid state particle concentrations when crystallizing the binary metallic melts in the case of microsystems have been stated by BAIKOV et al. (1980a, 1979b, 1985). A difference between the one-component and binary metallic melts' crystallization for both micro- and macrosystems is concluded in two-phase transitional zone's thermodynamics which for the binary metallic melts has been described in detail by BAIKOV et al. (1980a). A scheme of a two-phase transitional area for the binary alloys within a small supercooling region is analogous to one shown for one-component metallic systems in Figure 2, i.e. when the spontaneous fluctuations of the limited spectrum solid state particle concentrations acting, the above-mentioned zone has a step-like shape. As with the the one-component micro- and macrosystems for the binary micro- and macroscopic alloys between the n and N parameters in the case of given mechanism's action a well-known restriction is: $n \ll 1 + 1.4 \ln N$. Within a small supercooling region in the case of very dilute binary solid solution's crystallization for metallic microsystems the given mechanism leads to the normal law (see BAIKOV et al. 1979b, 1980b, 1983) (the below given quantity T_l is a liquidus temperature):

$$V = \frac{a}{\tau} \frac{q_{AA}}{kT\,T_l} \Phi(n, N, \theta_{AA})\, \Delta T\,,$$

where

$$\Phi(n, N, \theta_{AA}) = \frac{1}{n\sqrt{\pi}} \sum_{i=1}^{n} \sqrt{\bar{n}_i}\left[1 - \exp\left(-\frac{\bar{n}_i}{4}\right)\right] \Big/ \left| r + \left(\frac{1}{2}\right)^{i-1} [\theta_{AA} + r(\gamma_A^i)^2] \right| . \tag{4.1}$$

Here: $\bar{n}_i = N(\frac{1}{2})^{i-1} \left|r + (\frac{1}{2})^{i-1} [\theta_{AA} + r(\gamma_A^i)^2]\right|$, a — a mean interatomic distance of binary crystal considered; $\theta_{AA} = \frac{2}{3} q_{AA}/kT$; q_{AA} — the A-component melting heat per atom; $\gamma_A^i = (k_A^i)^{-1}$; $k_A^i = C_{1A}^i \mid C_{2A}^i$ — the A-component distribution coefficient of the (A + B) system; C_{1A}^i, C_{2A}^i — the A-component concentrations in an i-monolayer in crystalline and liquid states, respectively; $\Delta T = T_l - T$. The formula (4.1) is just for binary microsystems which have a well-developed transitional zone, i.e. when the condition $|\theta_{AA}/r| \ll 1$ is valid. Given fluctuation mechanism in the case of binary microsystems determines the two-phase transitional zone's shape by the $\{\bar{X}_{ii}\}$-set, where at $|\theta_{AA}/r| \ll 1$ and $\Delta T/T_l \ll 1$ conditions $\bar{X}_{ii} \cong (\frac{1}{2})^i$ ($i = 1, \ldots, n$). The step-like shape of the transitional zone is in good agreement with the available experimental evidence (see BORISOV et al. 1965, 1969; BORISOV, MATVEEV). When considering binary macrosystems (see BAIKOV, CHISTYAKOV; BAIKOV, ZELENEV) within a small supercooling region and for well-developed transitional zones the fluctuation mechanism in question in the case of very dilute binary solid solution's crystallization leads to normal kinetics with the following effective kinetic coefficient:

$$V^* = \lim_{N \to \infty} V = K^*_{\text{eff}} \, \Delta T \,,$$
$$K^*_{\text{eff}} = \frac{a}{\tau \, |r| \, n} \left[\sum_{i=2}^{n} \sum_{j=1}^{i-1} \frac{1}{1 + (\frac{1}{2})^{j-1} \, (k_A^j)^{-2}} \right] \frac{q_{AA}}{kT \, T_l} . \tag{4.2}$$

As seen from the formulas (4.1), (4.2), given fluctuation mechanism in the case of binary microsystems determines the normal crystallization kinetics by the N and n-parameters, for binary macrosystems the effective kinetic coefficient only depends on the n-parameter. In the cited works (BAIKOV, CHISTYAKOV; BAIKOV, ZELENEV) for some binary macrosystems on the base of In- and Ga metals crystallizing from the proper melts within a small supercooling region a comparison between fluctuation

Table 4
A comparison of theoretical and experimental data concerning the average crystallization rates of several binary alloys within a small supercooling region. Two-phase transitional zone's size is characterized by $n = 30$. The relative errors of experimental data $(\Delta V_{\text{exp}}/V_{\text{exp}})$ 100% for the given binary alloys are: 11.0%; 1.0%; 1.2%; 1.5%; 2.6%, respectively

Binary alloy	ΔT/K	V^*/mm · s^{-1}	V_{exp}/mm · s^{-1}
In + 9 at.% Bi	0.1	0.19	0.17
	0.3	0.57	0.51
	0.5	0.96	0.85
In + 4.6 at.% Pb	0.1	2.4	10.0
	0.3	7.2	30.0
In + 7.1 at.% Pb	0.1	2.5	2.6
	0.34	8.5	9.0
In + 9.5 at.% Pb	0.1	2.4	1.3
	0.8	19.0	11.0
Ga + 8 at.% Sn	0.5	7.5	19.0
	1.0	15.0	38.0

theory and experimental data has been presented. In Table 4 at small supercoolings some results concerning the very dilute binary solid solution's crystallization kinetics are shown. For all binary alloys on the base of In the spontaneous fluctuation periods estimated by virtue by LEAK, SWALIN; SWALIN 1959, 1961 are equal to $\tau = 10^{-11}$ s, in the case of binary alloy Ga + 8 at.% Sn the τ-value is 10^{-10} s. As seen from Table 4 the fluctuation theory of normal crystal growth based on the here considered fluctuation mechanism describes as a whole satisfactory available experimental evidence connected with binary melts' crystallization kinetics.

5. Conclusions

1. In the case of fluctuation mechanism of the "unlimited" spectrum solid state particle concentrations within a small supercooling region when crystallizing both the micro- and macroscopic one-component metallic melts the average crystal growth rate does not depend on two-phase transitional zone's size.
2. The mechanisms of the ordered "shortened" spontaneous fluctuations and of the spontaneous ones of the limited spectrum solid state particle concentrations in the case of the one-component metallic melt's crystallization within a small supercooling region lead to the same kinetic and effective kinetic coefficients, resvectively, for micro- and macrosystems. The given kinetic coefficients depend on the two-phase transitional zone's size. Both the above mentioned mechanisms give the step-like shape of the transitional zone. In this case there is a connection between the N- and n-parameters.
3. Within a small supercooling region in the cases of binary micro- and macrosystems the fluctuation mechanism of the limited spectrum solid state particle concentrations leads to the normal crystallization kinetics. In the case of very dilute binary solid solution's crystallization the effective kinetic coefficient depends on the two-phase transitional zone's size and on the A-component distribution coefficient of the (A+B) system.
4. In the cases of fluctuation mechanisms of the "unlimited" spectrum solid state particle concentrations and of the limited ones within a small supercooling region for some one-component and binary melts a comparison of theoretical and experimental data concerning the average crystallization rates has been presented. A good agreement between the fluctuation theory and experimental evidence has been established.

References

BAIKOV, Yu. A., CHISTYAKOV, Yu. D., MANEC, V.: Vereinigung für Kristallographie, Mitteilungen 1, 13. Jahrgang 1978, Berlin, p. 3–18 (a)

BAIKOV, Yu. A., ZELENEV, Yu. V., HAUBENREISSER, W.: phys. stat. sol. (a) **55** (1979) 123 (a)

BAIKOV, Yu. A., CHISTYAKOV, Yu. D., SCHNEIDER, H. G., RUTH, V.: Acta Metallurg. **29** (1981) 415; Mathem. Appendix to Acta Metallurg. **29** (1981) 1

BAIKOV, Yu. A., CHISTYAKOV, Yu. D.: Cryst. Res. Technol. **20** (1985) 1143 (a)

BAIKOV, Yu. A., CHISTYAKOV, Yu. D., MANEC, V.: Vereinigung für Kristallographie. Mitteilungen 2, 13. Jahrgang, 1978, Berlin, p. 3–18 (b)

BAIKOV, Yu. A., ZELENEV, Yu. V., HAUBENREISSER, W.: phys. stat. sol. (a) **59** (1980) K81 (a)

BAIKOV, Yu. A., ZELENEV, Yu. V., MOLOTKOV, A. P.: Kristall und Technik **14** (1979) 389 (b)

BAIKOV, Yu. A., CHISTYAKOV, Yu. D., SCHNEIDER, H. G., RUTH, V.: Cryst. Res. Technol. **18** (1983) 711

BAIKOV, Yu. A., ZELENEV, Yu. V., HAUBENREISSER, W.: phys. stat. sol. (a) **59** (1980) K155 (b)

BAIKOV, Yu. A., CHISTYAKOV, Yu. D.: Cryst. Res. Technol. **20** (1985) 1301 (b)

BAIKOV, Yu. A., ZELENEV, Yu. V.: Izv. AN SSSR, ser. Neorgan. mater. **22** (1986) 1656

BORISOV, V. T., GOLIKOV, I. N., MATVEEV, Yu. E.: Coll. works CNIIChM n. a. BARDIN, I. P. Issue 69, Moscow 1969, pp. 182–189

BORISOV, V. T., DUKHIN, A. I., MATVEEV, Yu. E., RAKHMANOVA, E. P.: Coll. Growth of Crystals **5**, Moscow 1965, pp. 68–71

BORISOV, V. T., MATVEEV, Yu. E.: Kristallographiya **6** (1971) 247
BORISOV, V. T.: Coll. works IV Allunion conference on crystal growth, Growth and Kinetics, Erevan 1972, pp. 25–29
GILMER, G., BENNEMA, P.: J. Cryst. Growth **13/14** (1972) 148; J. Appl. Physics **43** (1972) 1347
LEAK, V., SWALIN, R. A.: Trans. Met. Soc., AIME **230** (1964) 426
SWALIN, R. A.: Acta Metallurg. **7** (1959) 736
SWALIN, R. A.: Acta Metallurg. **9** (1961) 379

(Received, accepted November 4, 1987)

Authors' address:

c/o Prof. Dr. techn. Yu. D. CHISTYAKOV, Dr. phys. Yu. A. BAIKOV, Ing. M. V. AKULIONOK
Moscow Institute of Electronic Engineering
Avangardnaya 13
Moscow 125493, USSR

Book Review

J. PHILIBERT

Diffusion et Transport de Matière dans les Solides

Les Editions de Physique, Les Ulis 1985, 472 Seiten, zahlreiche Abbildungen und Tabellen, Sachwörterverzeichnis, Preis 250 F.F., ISBN 2-86883-004-8

Es gibt nur wenige Gebiete in der Festkörperphysik und der Festkörperchemie, die sowohl vom praktischen als auch theoretischen Standpunkt so bedeutend sind wie die Diffusion. Besonders die stürmische Entwicklung der Meß- und Rechentechnik im letzten Jahrzehnt gestattete ein tieferes Eindringen in die Elementarprozesse. Die größten Fortschritte sind bei Halbleitern und oxidischen Systemen zu verzeichnen. Trotz der Fülle an Monographien und Übersichten, die in der internationalen Literatur vorhanden sind, hat der Autor mit der Konzeption dieses Buches eine Art Marktlücke erschlossen: Er betrachtet die Diffusion als übergreifendes Phänomen. Hervorstechend ist die anschauliche Darstellung der Theorie und der mathematischen Bezüge, ohne daß die physikalische Bedeutung und die stofflichen und experimentellen Aspekte in den Hintergrund treten. Nach dem Studium des Buches hat der Leser einen Gesamtüberblick, einschließlich vieler Querverbindungen zu angrenzenden Gebieten wie z.B. der irreversiblen Thermodynamik und der Kristallographie. Um den Zusammenhang nicht zu stören, sind spezielle Probleme am Ende jedes Kapitels als Anmerkungen beigefügt.

Das Buch ist in zehn Hauptabschnitte untergliedert. Die ersten drei beinhalten allgemeine mathematische und physikalische Grundlagen sowie Diffusionsmechanismen. Daran anschließend wird die Selbstdiffusion behandelt, dabei reicht das Spektrum von Metallkristallen über Halbleiter und Ionenkristalle bis zu den Molekülkristallen. Ausführlich wird auf die verschiedenen Mechanismen der Teilchenbewegung im Festkörper eingegangen. Die Abhängigkeiten des Diffusionskoeffizienten von äußeren Parametern wie Druck, aber auch von inneren Kräften z.B. elektrischer Felder, werden beleuchtet. Letztere spielen insbesondere bei der Diffusion von Fremdatomen in Halbleitern eine Rolle, da die bereits vorhandene Untergrunddotierung das Diffusionsprofil erheblich beeinflussen kann. In den folgenden Kapiteln steht die Diffusion von Fremdatomen im Mittelpunkt. Neben allgemein akzeptierten Modellvorstellungen wurden aktuelle Beispiele aus der Literatur diskutiert wie der Stofftransport in Ionenleitern und amorphen Materialien, die Interdiffusion (als klassisches Beispiel der Kirkendall-Effekt) und die Darkenschen Gleichungen. Unter Diffusion in begrenzten Systemen diskutiert der Autor zwei Probleme — den Einfluß der Realstruktur und den Transport an der Oberfläche. Desweiteren erfolgt eine phänomenologische Betrachtung. Ausgehend von der Thermodynamik irreversibler Prozesse werden mit Hilfe phänomenologischer Beziehungen spezielle Diffusions- und Transportvorgänge im Festkörper beschrieben. Der Leser wird weiterhin mit Methoden und Verfahren zur experimentellen Bestimmung von Diffusionsprofilen sowie zur Untersuchung der vorherrschenden Phänomene bei der Teilchenbewegung vertraut gemacht. Schließlich wird in einem umfangreichen Kapitel gezeigt, daß zahlreiche physikalische und chemische Prozesse Diffusionsvorgänge beinhalten und durch sie zu erklären sind, wie z.B. die Oxidation von Metallen und binären Verbindungen.

Am Schluß jeden Abschnittes findet der Leser eine umfangreiche Bibliographie, unterteilt in allgemeine und spezielle Literatur.

Das Buch spricht einen breiten Leserkreis an und ist sicher sowohl für Spezialisten als auch für diejenigen, die sich allgemein informieren wollen, eine sehr nützliche Lektüre.

E. NOWAK

Cryst. Res. Technol.	23	1988	5	621–627

R. S. MADOYAN, O. A. KHACHATURYAN, S. G. GRIGORYAN, A. I. GABRIELYAN

Research Institute of Condensed Matter Physics, Yerevan State University
Yerevan, USSR

Crystallization Kinetics of Lithium Niobate Films from Limited Volume of Melt Solution

The growth kinetics of $LiNbO_3$ films from a limited volume of melt solution is observed. We developed crystallization models describing the character of mass transfer in the liquid phase for isothermal and non-isothermal epitaxy. Analytical expressions have been derived connecting the film thickness with the system growth parameters. The experimental values being in good agreement with the calculated ones.

Рассмотрена кинетика роста пленок ниобата лития из ограниченного объема раствора-расплава. Построены модели кристаллизации, описывающие характер процессов массопереноса в жидкой фазе для изотермического и неизотермического вариантов эпитаксии. Получены аналитические выражения, связывающие толщину пленки с параметрами ростовой системы, экспериментальные значения хорошо согласуются с расчетными.

Introduction

Liquid-phase epitaxy occupies a special place in the ferro-electric film technology (LINÉS, GLASS). The methods of film precipitation from a limited volume of melt solution, which allows to control the epitaxy process and to obtain layers of given parameters (PANISH, SUMSKI; KHACHATURYAN, MADOYAN) have immensely developed.

The present paper investigates crystallization kinetics of lithium niobate films from a solution in an Li_2O–V_2O_5 melt placed between two parallel substrates: the capillary technique (MADOYAN et al.). The layer deposition is carried out isothermally from the super-saturated solution and is due to cooling of the saturated solution. In the paper we attempted to describe the growth kinetics of lithium niobate films for both cases.

Nonisothermal method

Most of works on growth kinetics of epitaxial films from the melt solution are based on the diffusion approximation, when the film growth rate is limited by a mass transfer of the solved component to the crystallization front (MOON). This rather simplifies the solution of arising problems and allows to predict the film parameters in some cases. We, too, have considered the kinetics of nonisothermal growth of $LiNbO_3$ films in the diffusion approximation.

Figure 1 plots the capillar system with concentration distribution of the solved component in the melt. The gap between the substrates is much less than their dimensions. It corresponds to the real conditions and allows to proceed to the one-dimensional problem. No local supercooling is supposed to be present, the initial concentration C_0 being the same within the whole solution thickness.

Cooling of the growth system at constant rate α results in a supersaturation of the melt solution and precipitation on the substrate surfaces and formation of a concentration profile with maximum supersaturation in the liquid phase volume. In such

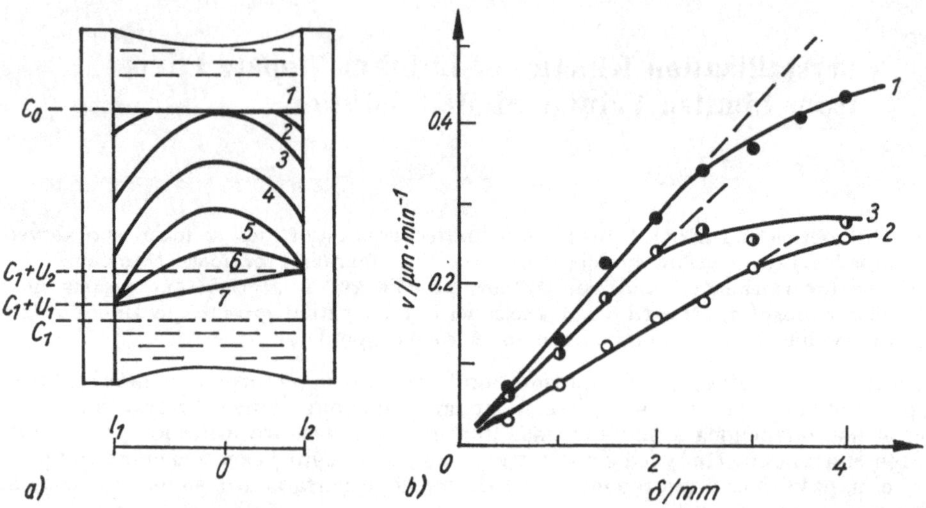

Fig. 1. Non-isothermal liquid phase epitaxy
a) Concentration distribution in the capillar system at sequential time moments (1 – 6).
b) Dependence of film precipitation rate on the magnitude of the capillar gap at different cooling rates: 1. $\alpha = 0.2$ grad · min^{-1}, 2. $\alpha = 0.4$ grad · min^{-1}, 3. $\alpha = 0.6$ grad · min^{-1}

systems, as a rule, the concentration at the crystallization front is considered to be in equilibrium at the given temperature (MOON). However, it cannot account for the difference in growth rates of different faces. In our model the $C(l)$ concentration at the substrate surface is supposed to exceed the equilibrium one C_1 by some characteristic value U, which depends on the substrate orientation. It leads to formation of an asymmetrical concentration profile with the maximum at $X = 0$ $\left((\partial C/\partial x) = 0\right)$, which, in general, is not in the capillar centre. Under stationary conditions $(Dt/l^2 \gg 1)$ this profile does not change and the precipitation on each substrate occurs from the liquid phase layer of thicknesses l_1 and l_2.

Using the solution of the diffusion equation (MOON) we obtain the expression for the concentration change in the region from the substrate surface to $X = 0$.

$$\frac{\partial C}{\partial x} = \frac{\alpha l_i}{mD}\left\{\frac{x}{l_i} - 2\sum_{n=0}^{\infty}\frac{(-1)^n}{\lambda_n^2}\exp\left[-\frac{\lambda_n^2 D(t - U_i m/\alpha)}{l_i^2}\right]\sin\left(\lambda_n\frac{x}{l_i}\right)\right\}, \tag{1}$$

where D is the diffusion coefficient, m the liquidus curve slope, $\lambda_n = (n + \frac{1}{2})\pi$; $i = 1, 2$.

From (1) it is easy to obtain the expression for maximum supersaturation in the liquid phase.

$$\Delta C_m = \frac{\alpha l_i^2}{2Dm} + U_i\,. \tag{2}$$

As well as for thicknesses l_1 and l_2 of the liquid phase layers, from which the precipitation on each substrate occurs

$$l_{1,2} = \frac{1}{2}\left(\delta \pm \frac{2DU_{21}m}{\alpha\delta}\right), \tag{3}$$

where $\delta = l_1 + l_2$ is the gap between the substrates, $U_{21} = U_2 - U_1$.

The increase of the gap or cooling velocity may lead to spontaneous crystallization in the liquid phase volume. When the substrates are similar, for the critical gap we have

$$\delta_{\text{cr}} = \left[2D\frac{m}{\alpha}(\Delta C_{m0} - U)\right]^{1/2}, \tag{4}$$

where ΔC_{m0} is the critical supersaturation of the melt solution. Along the capillar centre a new crystallization front appears, for which $U = 0$ may be accepted, since spontaneous non-oriented crystallization occurs. The effective thickness of the liquid phase, from which the epitaxial layer is being precipitated, becomes equal to

$$l_{\text{ef}} = \frac{1}{2}\left[\frac{\delta}{2} - \frac{4DUm}{\alpha\delta}\right]. \tag{5}$$

Definition of the epitaxial film thickness, which is in fact the only measurable film parameter, is an important task. Usually the thickness is defined by means of the phase diagramme, ignoring the mass transfer. For lithium niobate epitaxy the formula has the form (6).

$$h_0 = \frac{\varrho_1}{\varrho_s}\delta\frac{(C_0 - C_i)\,M_1}{C_1(M_1 - M_2) + M_2}, \tag{6}$$

where ϱ_l and ϱ_s are the melt solution and crystal densities, M_2 and M_1 are the molecular weights of the solvent and solved component, respectively. Note, that the same formula may be used for the isothermal epitaxy.

In real experiments the film thickness turn out to be less than the calculated ones, due to non-stationary processes.

The developed model allows to determine the main parameters of the crystallization system. The characteristic supersaturation U can easily be obtained from the difference between the calculated (6) and experimentally obtained film thickness under growth conditions, maximum close to equilibrum. For the diffusion coefficient from (3) we obtain:

$$D = \frac{\alpha\delta(l_1 - l_2)}{2U_{21}m}, \tag{7}$$

where l_1 and l_2 are determined by the film thickness with epitaxy on different substrates. To judge in experiments whether the built model corresponds to the real growth system, the main condition of the diffusion approximation must be satisfied, i.e. absence of convective mass transfer in the melt solution. The calculations show that in capillar systems the contribution of convective fluxes is negligible with small gaps and grows with the increase of (7). The linear dependence of the growth rate V on $(\delta \cdot \partial C/\partial t)$ witnesses the fact that the mass transfer, under these conditions, is limited by the molecular diffusion.

Deviation from non-linearity, when δ exceeds some value, is connected with the natural convection in the solution, which originates under gravitation conditions because of the density change in the melt solution volume.

Figure 1b shows the dependence of the film growth rate on the gap value. At small values of δ the dependence is of linear character. For $\alpha = 0.2$ deg min^{-1}, the contribution of the convective mass transfer is negligibly small up to $\delta^* \approx 3$ mm. For $\alpha = 0.4$ deg · min^{-1}, the value of δ^* decreases. It is obviously connected with the increase of temperature gradients and non-homogeneities of the liquid phase density in the melt solution. For $\alpha = 0.6$ deg · min^{-1} at $\delta \geqq 2$ mm, the layer growth rate (its thickness as well) does not change substantially, due to crystallization in the volume.

Under the conditions of diffusion mass transfer the developed model has shown close agreement with real experimental results for $LiNbO_3$ epitaxy on the substrates of $LiTaO_3$. We have obtained values for characteristic supersaturation at precipitation on the planes of 0001-Z and 1210-Y orientations: U_z 0.39 ± 0.05, mol.% and U_y = 0.24 ± 0.05 mol.%.

The capillar of Y, Z oriented substrates leads to precipitation of films of different thicknesses. When the cooling rate is 0.16 deg · min^{-1}, they may differ threefold. From the formula (7) for the diffusion coefficient we obtained $D \approx (1.5 \pm 0.7)\ 10^{-5}$ cm^2 × s^{-1}. For 20 mol.% of $LiNbO_3$ the maximum critical supersaturation ΔC_{m0} was calculated to be equal to 1.89 mol%. The calculations reveal the values of critical gaps 4.4, 4.0, and 2.8 mm for the values of α 0.16, 0.2, and 0.4 deg · min^{-1}, respectively. In contrast to rapid cooling ($\alpha = 0.6$ deg · min^{-1}, $\delta_{cr} = 2$ mm), at slow rates the convective mixing is prior to the crystallization in the volume, whereas the maximum concentration falls and critical supersaturation is not achieved.

Thus, under the conditions of diffusion mass transfer the developed model is in good agreement with the experimental results of lithium niobate epitaxy on $LiTaO_3$ substrates at forced cooling. Discrepancy between experimental and calculated values of film thickness in the diffusion regime does not exceed 10%.

Isothermal epitaxy

If in non-isothermal epitaxy the limiting processes are those which are determined by the system's cooling rate, in isothermal epitaxy the velocity of diffusion processes changes as film grows. So the problem of the relative contribution of diffusion and surface processes into the growth kinetics remains open. That is why we have considered the general case with due account of the kinetic coefficient. Isothermal epitaxy is observed under quasi-stationary conditions, which occur in the systems under consideration ($D/\delta V \gg 1$).

As in the case of non-isothermal epitaxy crystallization occurs from a limited volume, consisting of two similar parallel substrates, which are at a distance of δ, (Fig. 2a). The quantitative dependence of the film thickness on time at given initial supersaturation amounts to solving the differential equation, which describes the diffusion of the solved crystallizing material inside the capillar with corresponding initial and boundary conditions

$$\frac{\partial C}{\partial t} = \frac{2D}{\delta} \frac{\partial^2 C}{\partial y^2} + V \frac{\partial C}{\partial y}, \qquad 0 < y < 1 - \frac{2}{\delta} \int_0^t V\, dt\,, \tag{8}$$

where y is the dimensionless distance (in $\delta/2$ units), reckoned from the growth front. The change interval of y is caused by symmetry about the capillar centre. The initial and boundary conditions of the problem are:

1) at $t = 0$; $C(y; 0) = 0$

2) at the capillar centre $\frac{\partial C}{\partial y} = 0$ at $y = 1 - \frac{2}{\delta}\int\limits_0^t V\,dt$

3) at the growth front the condition of the crystallizing material mass conservation occurs:

$$D\frac{\varrho_l M_1}{\varrho_s[C_1(M_1 - M_2) + M_2]}\frac{\partial C}{\partial y}\bigg|_{y=0} = V\frac{\delta}{2}\left\{\frac{\varrho_l M_1}{\varrho_s[C_1(M_1 - M_2) + M_2]}C - 1\right\}\bigg|_{y=0}$$

where

$$V = q\frac{\varrho_l M_1}{\varrho_s[C_1(M_1 - M_2) + M_2]}(C - C_1)|_{y=0}\,.$$

q is the kinetic coefficient. The boundary conditions are written ignoring U_0, since in real systems $U_0 \ll \Delta C_0$.

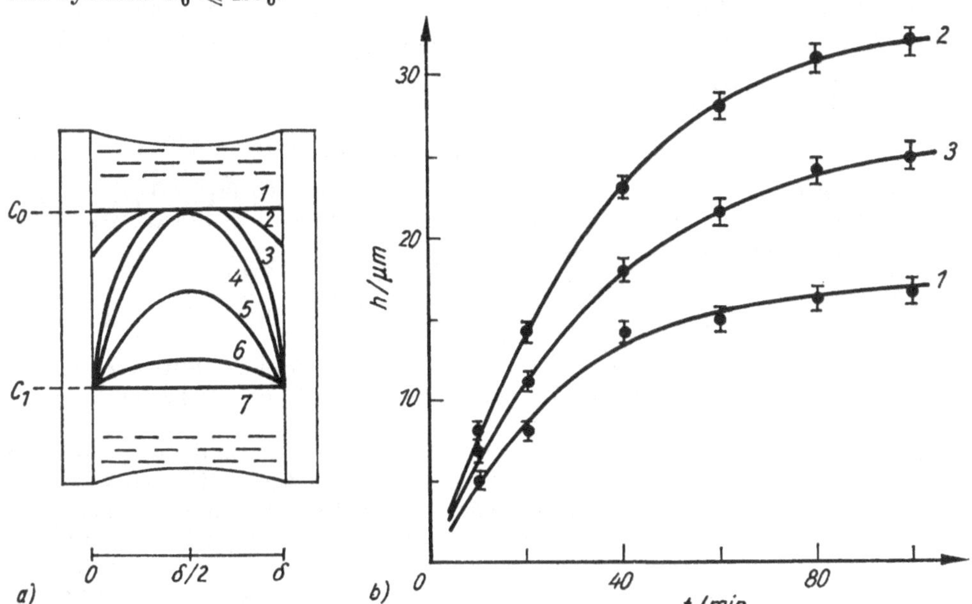

Fig. 2. Isothermal liquid phase epitaxy. a) Concentration distribution in the capillar system at sequential time moments (1 – 6), b) Dependence of film thickness on precipitation time at different initial supercooling values: 1. $\Delta T = 20°$, $\delta = 1.0$ mm, 2. $\Delta T = 40°$, $\delta = 1.0$ mm, 3. $\Delta T = 20°$, $\delta = 1.6$ mm

Under the conditions of liquid phase epitaxy the initial supersaturation $\Delta C_0 = C_0 - C_1$ is so negligibly small that the maximum film thickness (6) is much less than the capillar half-width.

Solving (8) by the method of separation of variables, we shall obtain the dependence of film thickness h on time:

$$h(t) = \int\limits_0^t V\,dt = \tau V_0 \sum_{n=1}^{\infty}\frac{2\gamma}{(\nu_n^2 + \gamma^2 + \gamma)\,\nu_n^2}[1 - \exp(-\nu_n^2 t/\tau)]\,, \tag{9}$$

where $\tau = \delta^2/4D$ is the diffusion time, $V_0 = V(t = 0)$,

$$\gamma = \frac{\delta q}{2D}\left\{1 - \frac{\varrho_1 M_1}{\varrho_s[C_1(M_1 - M_2) + M_2]} C_1\right\}$$

ν_n is the root of the equation $\gamma\nu = tg\nu$.

The first component in (9) does not depend on t. It is the final thickness of the film h_0.

At $\gamma \gg 1$ and $t \gg \tau/\pi$ one term may be left in the series (9)

$$h = h_0 - \frac{\tau V_0 2\gamma}{(\nu_1^2 + \gamma^2 + \gamma)\,\nu_1^2} \exp(-\nu_1^2 t/\tau)\,, \tag{10}$$

admitting a relative error

$$\frac{\Delta h}{h_0} < \frac{\nu_1^2 + \gamma^2 + \gamma}{6(\gamma^2 + \gamma)} \nu_1^2 \exp[-(\nu_2^2 - \nu_1^2)\, t/\tau]\,. \tag{11}$$

Note, that at $t = \tau$ the concentration profile change, caused by diffusion, reaches the point $\delta/2$ (Fig. 2a). At $t \ll \tau$ the film grows, as in an infinite capillar and the dependence $h(t \ll \tau)$ is determined by solving equation (8) with boundary conditions.

$$\begin{aligned} &\left.\frac{\partial\, \Delta C}{\partial y}\right|_0 - \gamma\, \Delta C|_0 = 0; \qquad \Delta C = C - C_1\,, \\ &\Delta C_1(z; 0) = C_0 - C_1; \qquad \Delta C(\infty; t) = C_0 - C_1\,. \end{aligned} \tag{12}$$

From (12) it is obvious that at $t \ll \tau_k = D^{-1}q\{1 - \varrho_1 M_1 \varrho_s^{-1}[C_1(M_1 - M_2) + M_2]\}^{-1}$ the growth kinetics regime occurs, i.e. the film growth rate is determined only by the kinetic coefficient and initial supersaturation

$$V = V_0 = q \frac{\varrho_1 M_1}{\varrho_s[C_1(M_1 - M_2) + M_2]} (C_0 - C_1)\,. \tag{13}$$

The equation (10) which is valid for $\gamma > 1$, $t \ll \tau/\pi$ describes the diffusion regime, where the growth rate is determined by the diffusion coefficient, the capillar width and is weakly affected by q.

Dependence of film thickness on time for different values of ΔC and δ was measured to find out the character of $LiNbO_3$ crystallization from the $LiNbO_3$–$LiVO_3$ melt solution. The measurement results are shown in Figure 2b. The final thickness h_0 is equal to the maximum film thickness obtained with the increase of time delay. The measurements were made at $h \gg 4\,\mu m$, since for smaller h, the error is comparable with the film thickness. The linear character of dependences $h(t)$ plotted on logarithmic coordinates witnesses the dominance of the diffusion growth regime at $h \gg 4\,\mu m$ i.e. the whole film practically grows in the diffusion regime. The experimental results allow to estimate the values in (10), as well as to determine how much this formula is in error. Thus, for the curve I (Fig. 2b)

$$V > 2 \cdot 10^{-6}\ \text{cm} \cdot \text{s}^{-1}, \varrho_1 M_1 \varrho_s^{-1}[C_1(M_1 - M_2) + M_2]^{-1} (C_0 - C_1) > 1.7 \cdot 10^{-2}$$

which corresponds to the kinetic coefficient $q > 1.2 \cdot 10^{-4}$ cm · s^{-1}. Eq. (10) shows the best agreement with the experimental results in accord with at $D = 0.5 \cdot 10^{-5}$ cm · s^{-1}, which corresponds to (7) in order. To estimate the error in using (10) we use (11), assuming $V > 2 \cdot 10^{-6}$ cm · s^{-1}, $\Delta C_0 = 10^{-2}$, $q > 2 \cdot 10^{-4}$ cm · s^{-1}, $\gamma > 2$, $\nu_2^2 - \nu_1^2 > (1 + \pi)\,\pi/2$. At $t/\tau > \pi^{-1}$ we obtained $\Delta h/h_0 < 10^{-1}$, which is within the error of measuring h. Thus in the growth system under consideration the kinetic regime occurs at $t \ll \tau_k \cong 5 \cdot 10^2$ s, whereas the diffusion one at $t \gg 2/\pi \cdot 10^3$ s.

This confirms the conclusion that in the film growth process the diffusion regime dominates.

Conclusion

A crystallization system model which completely describes the lithium niobate film precipitation in a capillar system at liquid phase epitaxy was developed on the basis of a series of investigations.

Good agreement between the calculated and experimental values allows to predict the film parameters and to control successively the crystallization process by means of the developed model.

References

KHACHATURYAN, O. A., MADOYAN, R. S.: Cryst. Res. Technol. **19** (1984) 461
LINÉS, M. E., GLASS, A. M.: Principles and Application of Ferroelectrics and Related Materials, Oxford 1977
MADOYAN, R. S., GABRIELYAN, A. I., KHACHATURYAN, O. A.: Crystallization Kinetics of $LiNbO_3$ Films from Limited Volume of Solution in Melt. VII Internat. Conf. on Cryst. Growth, Program and Abstract, FR Germany 1983
MADOYAN, R. S., KHACHATURYAN, O. A.: Izv. Acad. Nauk Armen. SSR, Physika **18** (1983) 351
MILVIDSKII, M. G., ORLOV, V. P., TSEPILEVICH, V. G.: Izv. Acad. Nauk inorgan. matter **16** (1980) 1159
MOON, R. T.: J. Cryst. Growth **27** (1974) 62
PANISH, M. B., SUMSKI, S.: J. Cryst. Growth **11** (1971) 101

(Received July 30, 1987;
accepted October 21, 1987)

Authors' address:

Dr. Dr. O. A. KHACHATURYAN, R. S. MADOYAN, S. G. GRIGORYAN, A. I. GABRIELYAN
Research Institute of Condensed Matter Physics
Yerevan State University
Shopron str. 21
375090 Yerevan, USSR

Book Review

P. Engel

Geometric Crystallography. An Axiomatic Introduction to Crystallography

D. Reidel Publ. Co., Dordrecht 1986. 266 p. with 53 figs. 140.00 Dfl./59.00 US $/41.95 £. ISBN 90-277-2339-7.

Die geometrische Kristallographie wird von Nebenfächlern heute oft noch als Teilgebiet betrachtet, das sich über fast einhundert Jahre hinweg nicht wesentlich weiterentwickelt habe. Das vorliegende Buch ist ein Zeugnis dagegen. Die Mathematisierung der Kristallographie hat inzwischen einen solchen Grad erreicht, daß die gewohnten Einführungsprozeduren nicht mehr zum Verständnis der Fachliteratur ausreichen und eine zusammenfassende Darstellung des anspruchsvollen Begriffsapparates gebraucht wird.

Der Autor wählt die axiomatische Form der Präsentation mit einer Folge von Definitionen, Theoremen und Beweisen, durch die der Problemkreis in sehr kompakter und begrifflich strenger Weise behandelt werden kann. Damit wird aber der Text für die Mehrzahl der Anwender kristallographischer Erkenntnisse nicht als Einführung zu empfehlen sein. Um diese abstrakte Formulierung würdigen zu können, bedarf es einer gewissen Vertrautheit mit den zentralen Begriffen wie Gitter, Symmetrieoperation, Punktgruppe, Raumgruppe, Punktkonfiguration usw., nicht zu reden von entsprechenden Grundlagen der Algebra. Ist sie vorhanden, wird der Leser das Buch gern zur Hand nehmen, um das notwendige Rüstzeug für eine Art höherer geometrischer Kristallographie zu erwerben, die z.B. bei der Anwendung des Computers auf diesem Gebiet unerläßlich ist.

Ein Vorzug des Titels ist auch die Berücksichtigung der Raumteilungen, die sonst selten in Lehrbüchern der geometrischen Kristallographie zu finden sind. Hier hat auch der Autor eigene Ergebnisse beigesteuert.

Zu begrüßen ist weiterhin die Einbeziehung von Räumen der Dimension > 3 und nicht zuletzt die historischen Notizen am Ende einiger Kapitel, weil sie auch auf die jüngere Entwicklung Rücksicht nehmen.

Überraschend ist die Inkonsequenz in der Bezeichnung von Punkt- und Raumgruppen. Weniger gebräuchliche Wortschöpfungen wie 'stabilizer' sind dagegen mehr ein Ausdruck der weltweiten Uneinigkeit in Bezug auf Bezeichnungen rund um einzelne Punktlagen.

Das Buch liegt als Computermanuskript vor, was offenbar nicht vor zahlreichen Schreibfehlern schützt, die aber leicht zu identifizieren sind. Insgesamt eine nützliche Lektüre für alle, die sich in die mathematische Kristallographie einarbeiten wollen.

P. Paufler

Cryst. Res. Technol.	23	1988	5	629–633

B. WOLF, A. ZEHE

Sektion Physik, Technische Universität Dresden

Inference of System Configurations on Thickness Homogeneity and Lateral Dopant Distribution of MBE Grown Films

The common technique to produce film thickness homogeneity and composition uniformity across the wafer in MBE-growth is a slow rotation of the substrate while growth proceeds. This however is not compatible with most in-growth analysis techniques. In this paper it is demonstrated, that an optimum uniformity without rotation can be achieved by an optimized system configuration.

Zur Erzeugung einer homogenen Schichtdicke und gleichmäßigen Zusammensetzung von MBE-gezüchteten Schichten entlang dem Durchmesser des Wafers wird das Substrat während des Abscheideprozesses langsam rotiert. Das allerdings ist unverträglich mit den meisten Analyseverfahren, die während des Schichtwachstums angewendet werden. In der vorliegenden Arbeit wird gezeigt, daß eine gute Gleichmäßigkeit bei stationärem Substrat durch eine optimierte Systemanordnung erreicht werden kann.

1. Introduction

The specific nature of beam generation and propagation under Knudsen conditions in Molecular Beam Epitaxy entails particular care in positioning the effusion cells and semiconductor wafers in order to ensure that flux nonuniformities are not mirrored in the epilayer thickness dimension and both dopant and mole-fraction homogeneities. To accomplish uniformities of better than one percent, wafers are generally rotated during growth (BACHRACH, KRUSOR; PARKER). This however is not compatible with most in-growth analysis techniques:

RHEED- and LEED-pattern strongly depend on the azimuth, and the analyzed spot by means of AES and XPS will permanently move outside the rotation axis. If complex in-growth analysis is desired (e.g. for research purposes), substrate rotation must be avoided. For this purpose an optimum homogeneity can be achieved by a defined substrate positioning beyond the crossover of the beam axii of the effusion cells, as it is shown below. Figure 1 displays schematically main instrumental components and their arrangement in a MBE research vessel with traditional thermal evaporation from effusion cells.

The practical need for high thickness and distribution homogeneity arises mainly from the following facts: (i) Quantum size effects, the working principle of so-called 'new physics' devices, depend sensitively on layer thickness and its lateral homogeneity (ZEHE, THESS; HERMAN), (ii) The possibility of epitaxially deposit GaAs on large (5″) Silicon wafers (BEAN) will have a considerable economic impact; But the problem of lateral homogeneity of composition becomes much more important than in case of ordinary III–V homoepitaxy; (iii) The prevailing philosophy in MBE (particularly Si-MBE) for industrial use has to be one of high yield. Identical mole-fractions of alloys over the whole diameter of the wafer determines cost-effectiveness and compatibility of MBE compared to more traditional technologies (LUSCHER et al.; KASPER). Thus, strong efforts have been made to improve MBE system configurations in order to produce optimum layer properties. In the present paper a special arrangement between wafer and effusion sources is discussed.

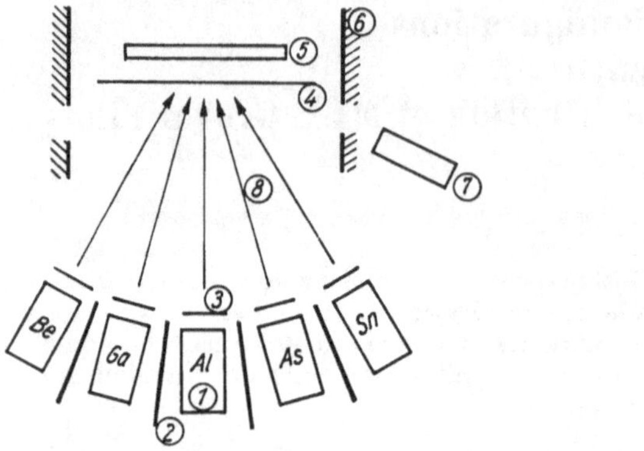

Fig. 1. Schematical arrangement of main instrumental components in MBE. (1) evaporation source, (2) vane baffle (thermal shielding), (3) individual beam shutters, (4) main shutter, (5) substrate (on heater block, mounted on a sample manipulator), (6) LN_2-cryo wall, (7) surface analyzing tool (AES, RHEED etc.), (8) molecular beams

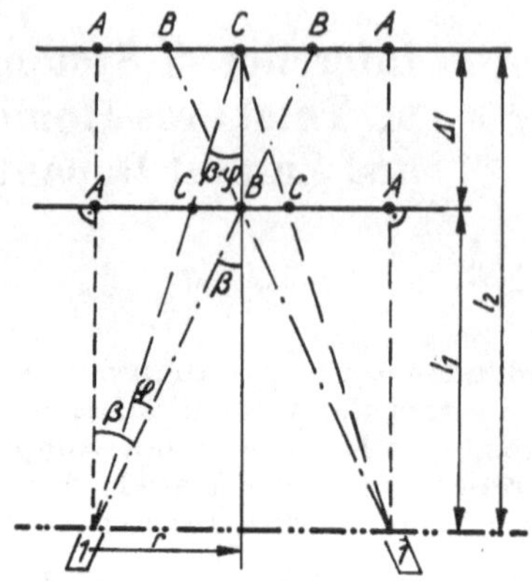

Fig. 2. Evaporation geometry in MBE; bold line: substrate position, dot dash line: geometrical axis of the effusion cell. A: thickness maximum of the isotropic point source, B: thickness maximum of the highly collimating source, C: thickness maximum of a real source, 1: effusion cell

2. General considerations

The flux angular density distribution of MBE effusion cells generates unavoidably a lateral inhomogeneity of constituents of the grown film accross the diameter of a wafer. At the extremal value (maximum, or even minimum in case of more than one effusion cell) of particle influx or layer thickness on the substrate a minimum lateral variation of thickness, doping level or mole-fraction is manifest. Let us consider two isotropic vapour sources (point sources). In this case each source will generate a thickness maximum at that point of the sample, which has the least distance to the source (point A in Figure 2). The position of these maxima will not be affected by vertically moving the substrate. Thus there is no coincidence of the individual distributions at finite distances.

The other extremum is a highly collimated stream, the flux density j of which is given by

$$j = \begin{cases} j_0\,, & \text{on the geometrical axis} \\ 0\,, & \text{off the geometrical axis}\,. \end{cases}$$

The location of thickness maxima of such sources depends on the vertical position of the substrate; and provided the substrate is fixed in the crossover of the geometrical beam-axii, the thickness maxima generated by the individual effusion cells will coincide.

In a practical situation the effusion cell has a certain angular distribution of j, different from both the isotropic and highly collimated case (VALYI).

Thus an optimum position of the substrate should be expected somewhere between the crossover of the beam-axii and infinity.

3. Evaluation of the optimum substrate position

According to Figure 2, β denotes the angle between the direction, perpendicular to the substrate, and the geometrical axis of the Knudsen-cell, and φ the polar-angle with respect to the cell's orifice. The layer thickness $d(\varphi)$ on the substrate writes (fig. 2)

$$d(\varphi) = f(\varphi) \cos^3 (\varphi + \beta) \quad (1)$$

with $f(\varphi)$ characterizing the normalized angular distribution of effusion density.

A minimum thickness variation requires

$$\frac{d}{d\varphi} d(\varphi) = 0 \,. \quad (2)$$

Under the assumption $f(\varphi) = \cos^n \varphi$ it follows:

$$\tan \varphi + \frac{3}{n} \tan (\varphi + \beta) = 0 \,. \quad (3)$$

A generalized expression for $f(\varphi)$ (according to DOBROWOLSKI) like

$$f(\varphi) = \sum_n (a_n \cos^n \varphi + b_n \sin^n \varphi) \quad (4)$$

requires the following equation to be fulfilled:

$$[\cos^{n-1} \varphi \, (- n a_n \tan \varphi - 3 a_n \tan (\varphi + \beta)) + \\ + \sin^{n-1} \varphi \, (n b_n - 3 b_n \tan \varphi \tan (\varphi + \beta))] = 0 \,. \quad (5)$$

Let us assume a term like $a_n \cos^n \varphi$ being dominant near the substrate, then (3) reduces to

$$\varphi = - \frac{3}{n + 3} \beta \quad (6)$$

small β and φ provided, that permit the following approximation:

$$\tan (\varphi + \beta) = \varphi + \beta \,. \quad (7)$$

Obviously (fig. 2) we have

$$\tan (\varphi + \beta) = \frac{r}{l_2} \approx \beta + \varphi \,, \quad (8)$$

$$\tan \beta = \frac{r}{l_1} \approx \beta \,, \quad (9)$$

which finally results in

$$\Delta l = l_2 - l_1 = \frac{3}{n} l_1 \,. \quad (10)$$

Δl represents the distance between the optimized substrate position, and the point, where it is commonly fixed (crossing of the cell axii).

4. Results and conclusions

As an example the following system parameters (according to Fig. 2) have been applied: Distance between cells $2r = 8$ cm, distance to the crossover $l_1 = 20$ cm (corresponding to an angle of inclination $\beta = 11.3°$), distance to the optimized substrate position $l_2 = 30$ cm.

The angular distribution of the cell is assumed to follow a $\cos^n \varphi$ — law, with $n = 6$ (determined by a selected length: diameter-ratio of the effusion channel).

Figure 3 shows the normalized lateral thickness distribution over the substrate, resulting from the two cells when the substrate is in crossover position.

In Figure 4 the total thickness distribution and the ratio d_A/d_B is given for this configuration, whereby d_A and d_B denote the thickness, grown from cell A and B,

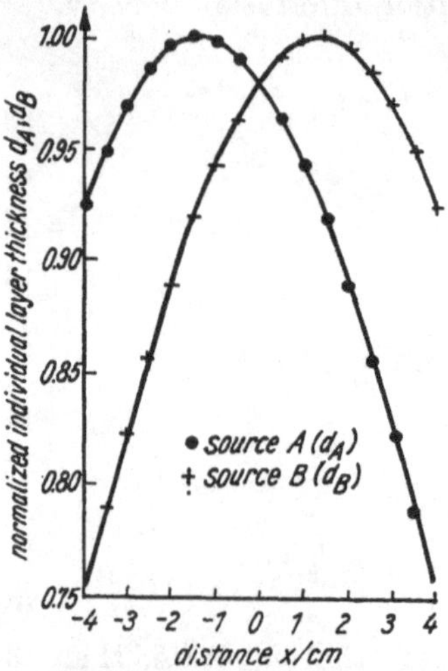

Fig. 3. Individual normalized thickness distributions generated by two identical effusion cells (substrate in the crossover of the geometrical beam axii, applied parameters see chapter 4)

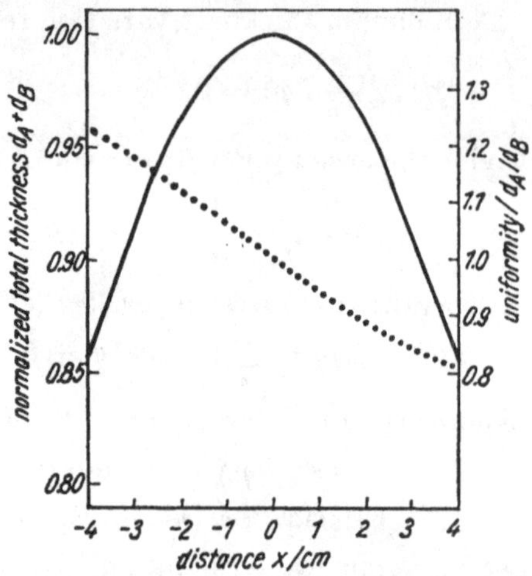

Fig. 4. Normalized total thickness distribution (bold line) and composition d_A/d_B (dotted line) for the same configuration as Figure 3

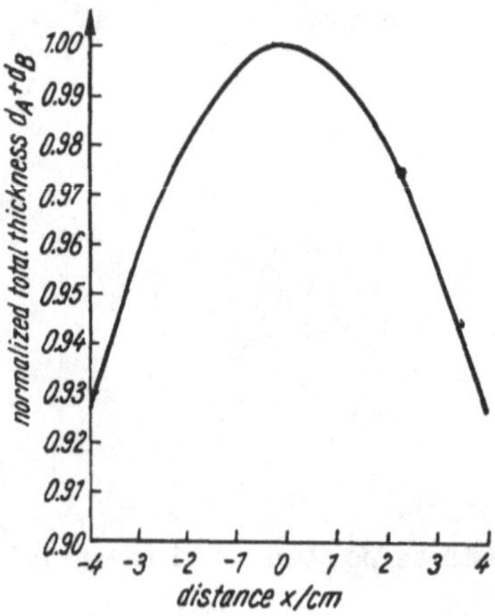

Fig. 5. Total thickness distribution for the substrate being in the optimized position ($d_A/d_B \equiv 1$ over the whole substrate)

respectively. For this example the region, displaying a nonuniformity in thickness of less than 3%, is about 3.4 cm in diameter, while the deviation in composition amounts to ±10%.

On the contrary, if the sample is positioned in a distance $l_2 = 30$ cm, resulting in a coincidence of the individual distributions of the two cells, a normalized thickness as shown in Figure 5 is achieved. The region characterized by a thickness variation less than 3% is extended to 5 cm. By taking the increased distance between cell and sample into account this is no considerable improvement, but the main advantage is the constant ratio d_A/d_B over the whole substrate. This is of major importance for the growth of ternary or more complex compounds, as e.g. $Ga_xIn_{1-x}As_yP_{1-y}$, where uniformity of the mole-fractions x and y over a large substrate is difficult to reach. The considerations above are valid in case of identical angular distribution of each cell. This can be achieved in one of the following ways: i) the use of decomposition sources (Gas Source MBE: GSMBE, as e.g. MOMBE), the geometry of which (that determines the angular distribution) can be choosen independently on the actual material (VALYI). ii) the use of Knudsen cells as dictated by its original meaning, where the effusion characteristic is independend on the filling level (VALYI); iii) the choice of geometrical measures of the evaporation cells according to the evaporation rates in case of quasi-Langmuir-evaporation in such a way, that the geometrical similarity of the effusion channels remains unaffected with sinking filling level during operation.

As a final conclusion, the use of just one source for a component cannot provide both thickness and compositional uniformity over large diameters under stationary substrate conditions.

References

BACHRACH, R. Z., KRUSOR, B. C.: J. Vac. Sci. Technol. **B1** (1983) 138
BEAN, J. C.: J. Cryst. Growth **82** (1987) 302
DOBROWOLSKI, G. M.: J. Vac. Sci. Technol. **A1** (1983) 1403
HERMAN, M.: Semiconductor Superlattices, Berlin 1986
KASPER, A.: Appl. Physics **A28** (1982) 129
LUSCHER, P. E., KNODLE, W. S., CHAI, Y.: Electronics, Aug. 28 (1980) 160
PARKER, E. C. H. (ed.): The Technology and Physics of MBE, New York 1985
VALYI, L.: Atom and Ion Sources, Budapest 1977
ZEHE, A., THESS, D. (ed.): Molekularstrahlepitaxie in der Siliciumgestützten Mikroelektronik-Technologie, TU Dresden 1986

(Received October 22, 1987;
accepted October 30, 1987)

Authors' address:

B. WOLF, Prof. Dr. Dr. h. c. A. ZEHE
Sektion Physik der Technischen Universität Dresden
Mommsenstr. 13
Dresden
DDR-8027

Book Review

M. Yussouff (ed.)

Electronic Band Structure and its Applications

Lecture Notes in Physics, vol. 283

Springer-Verlag, Berlin/Heidelberg 1987, 440 S., DM 73.–, ISBN 3-540-18098-2.

Der vorliegende Band der renomierten Springer Serie enthält 25 Vorträge, die von namhaften Wissenschaftlern auf einer internationalen Schule Ende 86 in Kanpur (Indien) gehalten wurden. Die Beiträge geben einen guten Überblick über den aktuellen Stand der Forschung auf dem Gebiet der Elektronenstruktur kristalliner und ungeordneter Festkörper. Das Buch enthält sowohl Beiträge zur Theorie als auch zu wichtigen experimentellen Methoden. Einige der Vorträge stellen eine gute Einführung in theoretische und experimentelle Standardmethoden dar mit ausreichenden Literaturhinweisen auf die Originalarbeiten. Das Buch kann daher allen Festkörperphysikern empfohlen werden, aber auch Physikstudenten mit einer festkörperphysikalischen Orientierung sollte der Band nützlich sein.

Auf einige Beiträge soll besonders hingewiesen werden. Mit der Dichtefunktionaltheorie beschäftigen sich die Vorträge von M. P. Das und G. Gunnarson u. a. Über erste Ergebnisse bei der Berechnung des Quasiteilchenspektrums im Rahmen der Dyson-Gleichung berichtet S. G. Louie. O. K. Andersen u. a. geben eine Einführung in die LMTO-Bandstrukturberechnungsmethode. Mit der Theorie Schwerer Fermionen und Quasikristallen beschäftigen sich die Beiträge von W. E. Pickett und J. B. Sokoloff. Einen Überblick über die Elektronentheorie ungeordneter Festkörper (Streutheorie, CPA) gibt A. Bansil. Von den experimentell orientierten Vorträgen seien die Beiträge über Photoelektronenspektroskopie (H. Neddermeyer), Compton-Streuung (G. P. Das) und Positron-Vernichtung (R. M. Singru) genannt.

W. John

Cryst. Res. Technol.	23	1988	5	635–639

Th. Kaufmann, G. Fuchs, M. Webert

Technische Hochschule Ilmenau
Sektion Physik und Technik Elektronischer Bauelemente, Ilmenau

Metalorganic Chemical Vapour Deposition of Oriented ZnO Films

Thin uniform films of ZnO could be obtained by MOCVD using $Zn(CH_3)_2$·THF adduct and H_2O as reactants. The film growth was carried out at atmospheric pressure. A vertical resistance-heated cold-wall reactor system was used. Highly *c*-axis oriented ZnO films being very smooth and having a low resistivity were grown on (100)Si and (100)Si/SiO_2 under certain conditions. Epitaxial growth could be achieved on $(1\bar{1}02)$-sapphire. The dependence of the properties of the films on the growth conditions was investigated.

Dünne ZnO-Schichten konnten durch MOCVD unter Verwendung von $Zn(CH_3)_2$-THF und H_2O als Reaktanten erhalten werden. Das Filmwachstum wurde bei Atmosphärendruck durchgeführt. Es wurde ein vertikales, widerstandsbeheiztes Kaltwandreaktorsystem verwendet. Auf (100)Si und (100)Si/SiO_2 wurden bei bestimmten Bedingungen sehr ebene ZnO-Schichten mit stark ausgeprägter *c*-Achsentextur abgeschieden. Epitaktisches Wachstum konnte auf $(1\bar{1}02)$ Saphir erreicht werden.
Die Abhängigkeit der Filmeigenschaften von den Wachstumsbedingungen wurden untersucht.

Introduction

Among the different chemical methods for producing ZnO films the use of zinc-organic compounds is becoming more and more important in the chemical vapour deposition (MOCVD). The procedures having been described in literature since 1980 are principally based on the reaction of dimethylzinc (DMZ) and diethylzinc (DEZ) with various oxidizing agents at a temperature between 450 and 800 K according to the formula

$$\begin{array}{c} \text{DMZ} \\ \text{DEZ} \end{array} + \begin{array}{c|c|l} O_2 & CO_2 & \text{organic com-} \\ H_2O & NO_2 & \text{pounds containing} \\ & N_2O & \text{oxygen} \end{array} \rightarrow \text{ZnO} + \text{“}x_r\text{”}$$

with x_r being a gaseous residue.

The direct reaction of DMZ and DEZ with O_2 or H_2O does not yield satisfying results (Shealy et al.; Roth et al.; Ghandhi et al.; Smith). There are occuring homogeneous gas reactions before the formation of the film in connection with tiny ZnO particles falling down onto the substrate and with a parasitic deposition at the reactor walls. On the one hand, the separate leading-in of the initial components up to the close vicinity of the surface of the substrate will reduce the homogeneous reaction. On the other hand, however, this will lead to an inhomogeneous deposition of the film due to an incomplete mixing of the gaseous initial components. The deposition at a lower pressure will only slightly improve the quality of the film (Roth).

The use of oxidizing agents being less reactive, i.e. CO_2, NO_2, N_2O, will suppress the homogeneous reaction, especially in case of lower pressure (LAU et al.). If there is no excitation by plasma (SHIMIZU et al.) or ultraviolet light (PETERS; SOLANKI; COLLINS) simultaneously with the thermal excitation, however, this will result in very low deposition rates.

The reactions of DMZ and DEZ with organic compounds containing oxygen proved to be suited for avoiding the said difficulties, with the deposition rates being satisfactory (WRIGHT et al.; ODA et al.; VIJAYAKUMAR et al.). There is no clear reference in literature to the application of adducts being used extensively for the deposition of III–V compounds and II–VI compounds (MOSS; ZAOUK; YASUDA et al.; MITSUHASHI et al.) in case of the MOCVD of ZnO (STORCH). The adducts of DMZ and DEZ with tetrahydrofuran (THF) seem to be suited for this application.

Experimental apparatus and procedure

The core of the laboratory system is the reactor box containing the thermostated bubbler made of stainless steel for the DMZ-THF adduct, THF, H_2O, and the dopant, as well as the vertical reactor (Fig. 1). This reactor consists of an upper part made of quartz which is cooled by water (internal diameter 100 mm) and a lower part made of stainless steel

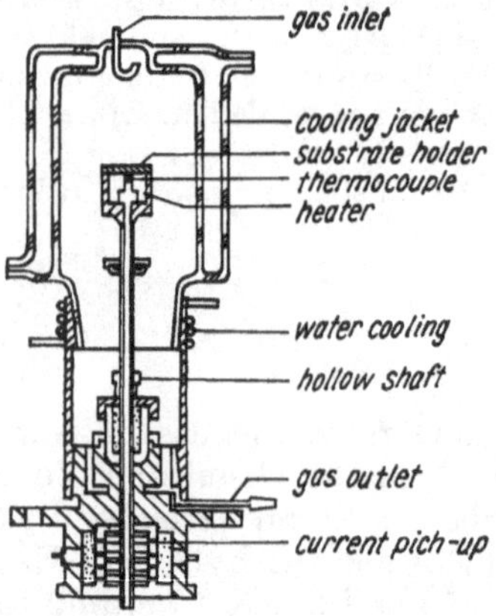

Fig. 1. The vertical MOCVD reactor

with a rotatable resistance-heated susceptor disc for substrates having a diameter of up to 75 mm. Two NiCr–Ni thermocouples and four filament current leads being electrically insulated are passed through the hollow shaft from the susceptor disc to the current pick-up. The susceptor disc is rotated via the hollow shaft by means of a coupling by the drive shaft of the motor. The shaft is sealed by a packing gland.

The temperatures between 573 and 773 K necessary for the deposition are reached after about 10 min and adjusted exactly to ± 1 K.

According to literature (THIELE), the DMZ-THF adduct, when boiling, underlies a partial, probably step-wise dissociation in the gasphase according to the formulae

$$(CH_3)_2Zn \times 2\,THF \rightleftharpoons (CH_3)_2Zn \times THF + THF$$

$$(CH_3)_2Zn \times THF \rightleftharpoons (CH_3)_2Zn + THF$$

In case of a slow fractional distillation there is evaporating first a higher proportion of free THF than of DMZ. Repeated distilling provides an adduct of the constant composition DMZ × 1.6 THF. According to these data, we obtained an adduct containing only 1.6 mole THF per mole DMZ already after the second very slow distilling via a column. We were able to show that an adduct of this composition during the saturation of a carrier gas flow will enter the gasphase unchanged at a temperature of 293—298 K, which is the working range of the bubbler in the deposition of films.

Results and discussion

ZnO films were grown on various kinds of substrates. The temperature was varied from 520 to 770 K. At the lower end of the temperature range we obtained amorphous layers. It is obvious that at temperatures higher than 770 K homogeneous gasphase reactions occur to an increasing extent resulting both in the formation of non-uniform films and the deposition of white powders on the reactor walls.

During the first stage of our work by the use of DMZ-THF adduct and THF alone, the properties of the ZnO films were considerably scattered. The growth rate was small. When a small amount of water was added as oxidizing agent, the scattering of the properties of the films disappeared and the growth rate increased. The growth conditions employed in the present work are listed in Table 1.

Table 1
Growth conditions

Substrates	(100)Si, (100)Si/SiO_2, $(1\bar{1}02)$-Al_2O_3, glass
Substrate temperature/K	570—750
Total gas flow (N_2, He/$cm^3 \cdot min^{-1}$)	3000 ... 4500
DMZ-THF adduct flow rate/mol · min^{-1}	$1 \cdot 10^{-5}$ ... $4 \cdot 10^{-5}$
THF-flow rate/mol · min^{-1}	$5 \cdot 10^{-5}$ ... $1.5 \cdot 10^{-3}$
H_2O-flow rate/mol · min^{-1}	$7 \cdot 10^{-6}$ ... $2 \cdot 10^{-4}$

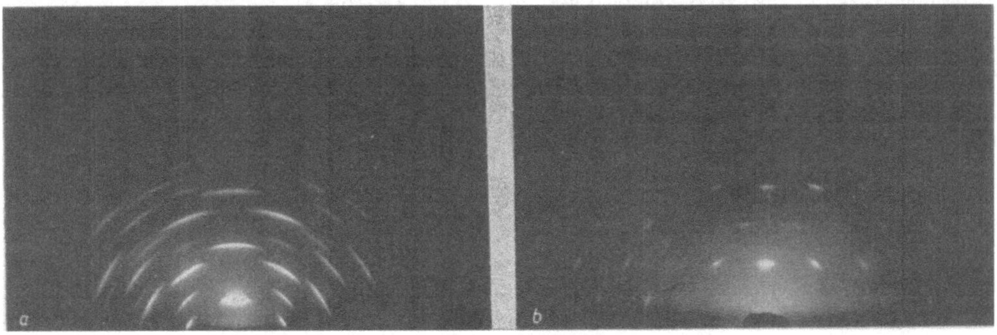

Fig. 2. RHEED diagram of ZnO films grown on (100)Si/SiO_2 (a) and $(1\bar{1}02)Al_2O_3$ (b)

The investigations concerning the structure of the ZnO films carried out by RHEED as well as X-ray diffraction techniques revealed that textured layers are formed on silicon and glass over the examined substrate temperature range. Highly *c*-axis-oriented ZnO films are obtained on (100)Si and (100)Si/SiO_2 at 650 to 750 K (Fig. 2a). Epitaxial growth could be achieved on $(1\bar{1}02)$ oriented sapphire (Fig. 2b).

In the examined range of substrate temperature, the decreasing THF/DMZ gasphase ratio ($\lesssim$ 100) results both in the formation of poorly textured films with decreasing growth rate and in the deposition of ZnO on the reactor walls. As mentioned above, a similar behavior occurs at a substrate temperature higher than 770 K.

Obviously, in both cases there takes place an enhanced decomposition of the DMZ-THF adduct forming free DMZ in the gasphase and therefore causing homogeneous gasphase reactions.

Using the growth conditions shown in Table 1, the growth rate of the ZnO films is only slightly dependent on substrate temperature (Fig. 3).

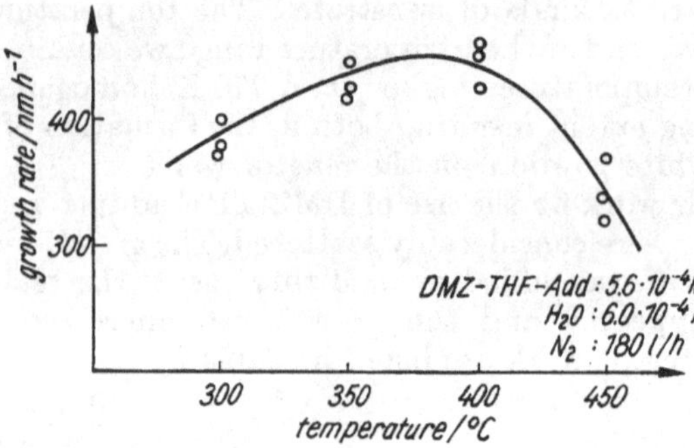

Fig. 3. Growth rate of ZnO films on Si/SiO_2 as a function of substrate temperature

The Auger analysis showed evidence of the presence of carbon in the ZnO films. The carbon content especially at the interface increased with increasing deposition temperature. The films obviously were free of carbon, if special deposition conditions were used, i.e. without THF or in case of the H_2O/DMZ-THF adduct gasphase ratio being greater than 5.

The layers grown on (100)Si (with and without SiO_2) are highly uniform and have extremely smooth surfaces under certain conditions. In layers grown on sapphire, the surface has a greater roughness. The differences in the thickness of layers on silicon wafers (50 mm in diameter) are very small (3–5%).

The ZnO films have low resistivities of 0.1–50 Ω cm. The electrical characteristics of layers are insignificantly affected by the growth conditions. The role of growth temperature and gasphase composition in controlling the resistivity of ZnO films still remains to be investigated.

Conclusions

Thin films of ZnO could be obtained by MOCVD using DMZ-THF adduct and H_2O as reactants. Because of its potential application for surface acoustic-wave devices, silicon was mostly used as the substrate. Highly *c*-axis oriented and very smooth ZnO films were grown at a temperature between 623–773 K on (100)Si/SiO_2 with a resistivity of 0.1–50 Ω cm. The problem of premature reactions, which often occurs in the MOCVD of ZnO, has been overcome in the mentioned temperature range suitable for the deposition by the use of DMZ-THF adduct and with a THF/DMZ gasphase ratio greater than 100.

Methods for increasing the resistivity (by doping) and the growth rate (by varying the growth conditions and also by the use of DEZ-THF adduct) are under investigation and will be described in detail in other papers.

The author's gratitude is due to Professor K.-H. Thiele of the Institute of Technology of Leuna-Merseburg for supplying the DMZ-THF adduct.

References

GHANDHI, S. K., FIELD, R. J.: Appl. Phys. Lett. **37** (1980) 449
LAU, C. K., TIKU, S. K., LAKIN, K. M.: J. Electrochem. Soc. **127** (1980) 1843
MITSUHASHI, H., MITSUI, W., KUKIMOTO, K.: J. Cryst. Growth **77** (1986) 485
MOSS, R. H.: J. Cryst. Growth **68** (1984) 78
ODA, S., TOKUNAJA, H., KITAJIMA, N., HANNA, J., SHIMIZU, J., KOKADO, H.: Jap. J. Appl. Physics **24** (1985) 1607
PETERS, J. W.: Patentschrift WO 84/00178
ROTH, A. P., WILLIAMS, D. F.: J. Appl. Physics **52** (1981) 6685
SHEALY, J. R., BALIGA, B. J., FIELD, R. J., GHANDHI, S. K.: J. Electrochem. Soc. **128** (1981) 558
SHIMIZU, M., MATSUEDA, Y., SHIOSAKI, T., KAWABATA, A.: J. Cryst. Growth **71** (1985) 209
SMITH, F. T.: Appl. Phys. Lett. **43** (1983) 1108
SOLANKI, R., COLLINS, E. J.: Appl. Phys. Lett. **42** (1983) 662
STORCH, W.: 4th Working Conference "Physics of semiconductor surfaces", Binz 1973, pp. 137–143
THIELE, K. H.: Z. anorg. allg. Chem. **319** (1962) 183
VIJAYAKUMAR, P. S., BLAKER, K. A., WIETING, R. D., WONG, B.: Patentschrift EP 0204563 A2
WRIGHT, P. J., GRIFFITS, R. J., COCKAYNE, B.: J. Cryst. Growth **66** (1984) 26
ZAOUK, A., SALVETAT, E., SAKAYA, J., MAURY, F., CONSTANT, E.: J. Cryst. Growth **55** (1981) 135

(Received, accepted October 29, 1987)

Author's address:

Dr. Th. KAUFMANN, Dipl.-Chem. E. FUCHS, Dipl.-Ing. M. WEBERT
Sektion Physik und Technik elektronischer Bauelemente der
Technischen Hochschule Ilmenau
PSF 327
Ilmenau
DDR-6300

Cryst. Res. Technol.	23	1988	5	641–645

P. PESHEV, S. DELINESHEV, V. PETROV, N. MANUILOV, P. YANKOV, I. BACHVAROV

Institute of General and Inorganic Chemistry, Bulgarian Academy of Sciences, Sofia, Bulgaria, and
Faculty of Physics, Sofia University
Sofia, Bulgaria

Bridgman-Stockbarger Growth and Spectral Characteristics of Al_2O_3:Ti^{3+} Single Crystals

Single crystals of Al_2O_3:Ti^{3+} (Ti ranging from 0.1–1.5 at.%) are grown by the Bridgman-Stockbarger method from a mixture of Al_2O_3 and Ti_2O_3 in a medium of 90% Ar and 10% H_2. The radial distribution of Ti^{3+} in the crystals is determined by measuring the optical transmission. The absorption spectra of the crystals and the luminescence spectra of Ti^{3+} in the Al_2O_3 matrix are obtained and the crystals fluorescence lifetime is determined. The optical characteristics of the crystals have shown good agreement with the corresponding characteristics of Al_2O_3:Ti^{3+} crystals grown by other methods.

Из расплава смеси Al_2O_3 и Ti_2O_3 методом Бриджмана-Стокбаргера в атмосфере из 90% Ar и 10% H_2 выращены монокристаллы Al_2O_3:Ti^{3+} с концентрацией титана в смеси от 0,1 до 1,5 ат.%. Радиальное распределение Ti^{3+} в кристаллах определено методом измерения оптической пропускаемости. Получены спектры поглощения кристаллов, спектры люминесценции йонов Ti^{3+} в матрице Al_2O_3 и определено флуоресцентное время жизни. Установлено, что оптические характеристики кристаллов находятся в хорошем соответствии с соответствующими характеристиками кристаллов Al_2O_3:Ti^{3+} выращенных другими методами.

1. Introduction

It was for the first time reported in 1982 (MOULTON) of the use of trivalent titanium-doped sapphire single crystals (Al_2O_3:Ti^{3+}) as a new active material for tunable lasers. Since then a significant num ber of investigations on this material, most of them dealing with its optical characteristics, appeared in the literature.

The single crystals of Al_2O_3:Ti^{3+} used for preparing laser hosts and the sample of which the optical properties were investigated, were grown by several methods: (1) Czochralski technique (MOULTON 1986; ALBERS et al.; LACOVARA et al.; POWELL et al.; SEVAST'YANOV et al.; KRUGLIK et al.; KAMINSKII et al.); (2) Verneuil method (ALBERS et al.; LACOVARA et al.; SEVAST'YANOV et al.; KRUGLIK et al.; KAMINSKII); (3) heat exchanger method (SCHMID, KHATTAK; LACOVARA et al.; MOULTON 1986); (4) vertical solidification of the melt, VSOM (POWELL et al.); (5) vertical gradient-freeze technique (SANCHEZ et al.) and (6) horizontal directed crystallisation (SEVAST'YANOV et al.).

The authors of the present paper developed a method (DELINESHEV et al.) for the preparation of single crystals doped with different transition metal ions (including Ti^{3+}). This method is based on the Bridgman-Stockbarger technique. The purpose of the present paper was: (1) to apply the new method to the growth of Al_2O_3:Ti^{3+}

crystals with various concentration of the dopant, titanium being introduced as Ti_2O_3; (2) to determine the optical characteristics of some of the crystals grown and (3) to compare the results obtained with data from the literature on single crystals grown by other methods.

2. Growth of $Al_2O_3 : Ti^{3+}$ single crystals

The experiments on single crystals growth were performed in cylindrical molybdenum containers using a GRANAT-2M apparatus in which the heating kit (consisting of a resistive tungsten heater and tungsten and molybdenum screens) was altered according to a previous work of the authors (PETROV et al.). The initial substances used for preparing the crystals were Al_2O_3 (99.999%, obtained in one of our laboratories) and Ti_2O_3 synthesized by a solid phase reaction of TiO_2 (spectrographically pure, Johnson, Matthey) with titanium sponge, its impurity content being 10^{-3}% (Koch & Light, Ltd). Single crystals were grown by both spontaneous nucleation and growth on a seed with a preliminary chosen orientation. The second method is preferable because it allows obtaining crystals with a preset orientation as well as better use of the initial material.

The initial Al_2O_3 was powdery alumina or sapphire crackle. The Al_2O_3 was mixed with Ti_2O_3 in the desired ratio, then the mixture was placed in a molybdenum container after removing oxide and organic contaminants from it and was heated for 1—4 hours at a temperature exceeding its melting point by 100—250 K. After that, the container with the melt was gradually lowered into the cold zone of the apparatus with a rate of 4—10 mm/h. The process took place under a gas mixture of argon and hydrogen in a 9:1 ratio at an overpressure of 0.02 MPa. Single crystals of $Al_2O_3:Ti^{3+}$ with a diameter of 10—14 mm and a length up to 150 mm were grown by the above method, the titanium concentration in the initial mixture ranging from 0.1 to 1.5 at.%. The colour of the crystals obtained changed from light pink to dark brown depending on the Ti concentration. Samples prepared from initial oxide mixtures containing more than 1.5 at.% Ti were very dark, opaque with respect to visible light and contained a lot of gas inclusions. Figure 1 presents crystals with different titanium contents.

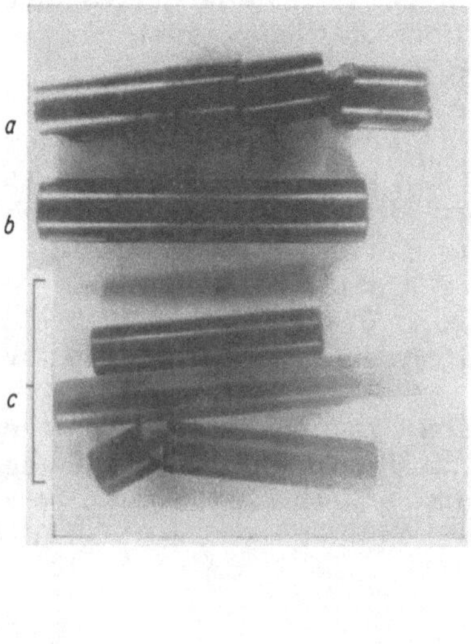

Fig. 1. As-grown and processed $Al_2O_3:Ti^{3+}$ single crystals with various Ti concentration in the initial material: (a) 2.0 at.%; (b) 0.4 at.% and (c) processed single crystals

3. Optical characteristics of Al_2O_3: Ti^{3+} crystals

The optical characteristics of the crystals were determined on cylindrical samples with different dopant concentrations and different sizes.

The measurement of the optical transmission of the crystals allowed determination of the distribution of Ti in them. These measurements were carried out with a wave length $\lambda = 532$ nm (a second harmonic of a YAG:Nd^{3+} laser) for which the absorption coefficient was $\alpha = 0.58\ cm^{-1}$ (SANCHEZ et al.). Figure 2 shows the distribution of

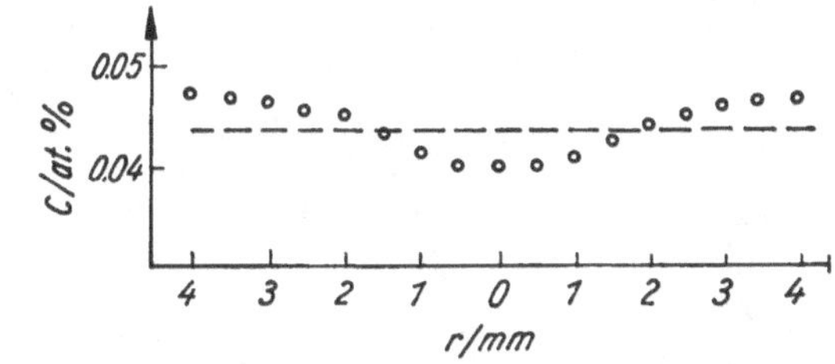

Fig. 2. Radial distribution of titanium in an Al_2O_3:Ti^{3+} single crystal obtained from an initial material containing 0.4 at. % Ti (d = 8 mm; L = 12 mm)

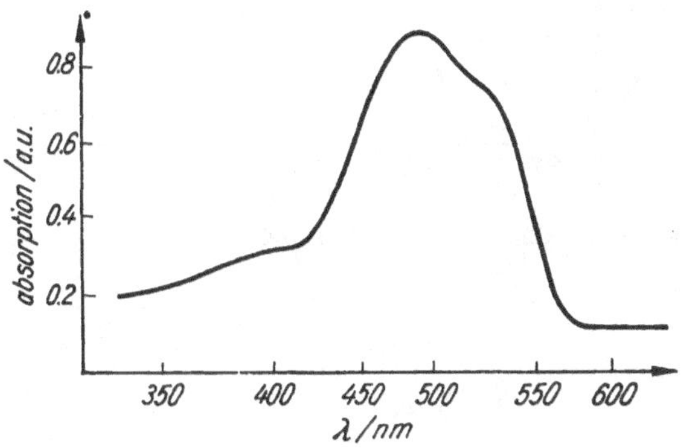

Fig. 3. Room temperature absorption spectrum of an Al_2O_3:Ti^{3+} single crystal obtained from an initial material with 0.4 at. % Ti

Ti^{3+} in a sample obtained from an initial material containing 0.4 at.% titanium for which the highest concentration gradient was observed. The mean value of the titanium concentration determined in the crystal was about 0.042 at.%, i.e. 10.5% of the concentration of Ti in oxides introduced into the initial mixture. This value exceeds those obtained by other growth methods where the highest value is not more than 7% (MOULTON 1986; POWELL et al.; ALBRECHT et al.).

Figure 3 presents the absorption spectrum of a single crystal with the same composition as that described above. Obviously, Ti^{3+} shows a relatively large absorption band which is asymmetrical and has a maximum at 490 nm. The asymmetry can be ascribed to overlapping of two absorption bands (with maxima at 490 and 530 nm, respectively) caused by the character of the $^2T_2 \rightarrow {}^2E$ energy transfer (SEVAST'YANOV et al.; KAMINSKII). The absorption spectra obtained in the present paper are in very good agreement with those for samples prepared by other methods (MOULTON 1986; ALBERS et al.; KRUGLIK et al.; KAMINSKII; SANCHEZ et al.; ALBRECHT et al.).

The luminescence spectrum of Ti^{3+} in the sapphire matrix (Fig. 4) was determined using the radiation of YAG:Nd^{3+} laser (second harmonic $\lambda = 532$ nm). Evidently, Ti^{3+} has a broad intense luminescence band ranging from 650 to 1000 nm (by the level 0.1) and exhibiting a maximum at 790 nm. The fluorescence lifetime for most samples was $\tau = 3.5\ \mu s$, which is in excellent agreement with the results obtained by other authors (MOULTON 1986; ALBERS et al.; POWELL et al.; KRUGLIK et al.; KAMINSKII; ALBRECHT et al.).

The internal transmission in the luminescence region (600—1000 nm) for samples with different Ti^{3+} concentration was also measured. Figure 5 shows the results for two of the samples. Evidently, with higher Ti^{3+} concentrations in the crystal, the internal transmission is weaker and shows a trend to decreasing in the long-wave region. This may be due, according to some authors (POWELL et al.), to charge transfer transition between coupled iron-titanium ion pairs $Fe^{3+}Ti^{3+}$–$Fe^{2+}Ti^{4+}$. Iron can be present in a negligible amount in Ti_2O_3. However, the formation of Ti^{4+} cannot be satisfactorily explained for the present (POWELL et al.).

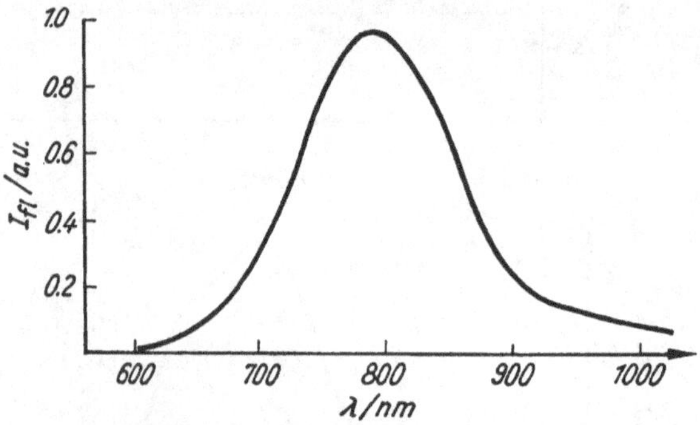

Fig. 4. Luminescence of Ti^{3+} in Al_2O_3

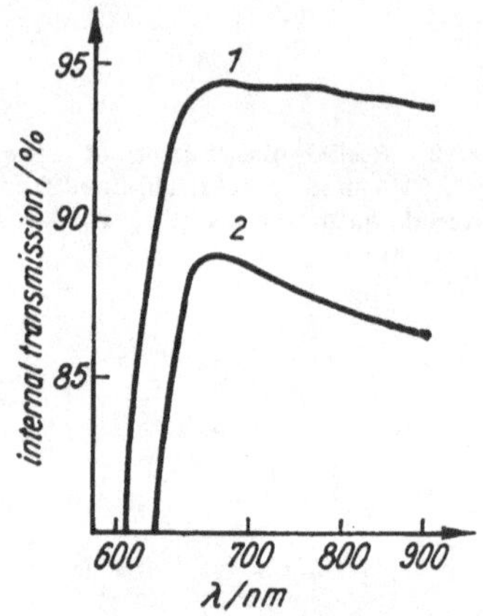

Fig. 5. Internal transmission in the luminescence region for samples with different Ti concentrations in the initial material: (1) 0.2 at. %; (2) 0.4 at. %

4. Conclusions

The results of the present paper permit the conclusion that the Bridgman-Stockbarger method may be used for growing sapphire single crystals containing ions of trivalent titanium as dopants. The properties of the crystals obtained make them appropriate for preparation of tunable laser hosts.

References

ALBERS, P., STARK, E., HUBER, G.: J. Opt. Soc. Amer. B **3** (1986) 134

ALBRECHT, G. F., EGGLESTON, J. M., EWING, J. J.: Opt. Commun. **52** (1985) 401

DELINESHEV, SV., BLIZNAKOV, G., PESHEV, P., PETROV, V., GYUROV, G., NIKOLOV, V., MANUILOV, N., KHUBANOV, Kh., GEORGIEV, G., YANAKIEV, A., ILIEV, K.: Bul. Pat. Appl. No 75204/04. 06. 1986

KAMINSKII, A. A. (ed.): Physics and Spectroscopy of Laser Crystals, Moscow 1986 (in Russian)

KRUGLIK, G. S., SKRIPKO, G. A., SHKADAREVICH, A. P., KONDRATYUK, N. V., ZHDANOV, E. A., ZABAZNOV, A. M.: Zh Prikl. Spektroskopii **42** (1985) 126

LACOVARA, P., ESTEROVITZ, L., KOKTA, M.: IEEE J. Quant. Electr. QE-**21** (1985) 1614

MOULTON, P. F.: Solid State Res. Rep., Lincoln Lab., Mass. Inst. Technol. (1982) pp. 15 to 20

MOULTON, P. F.: J. Opt. Soc. Amer. B **3** (1986) 125

PETROV, V., PESHEV, P., KHUBANOV, Kh., GEORGIEV, G., YANAKIEV, A., MANUILOV, N.: Bul. Pat. Appl. No 71035/10. 07. 1985

POWELL, R. C., CASLAVSKY, J. L., ALSHIEB, Z., BOWEN, J. M.: J. Appl. Physics **58** (1985) 2331
SANCHEZ, A., FAHEY, R. E., STRAUSS, A. J., AGGARWAL, R. L.: Opt. Lett. **11** (1986) 363
SCHMIDT, F., KHATTAK, C. P.: Laser Focus **19** (1983) 147
SEVAST'YANOV, B. K., BAGDASAROV, Kh. S., FEDOROV, E. A., SEMENOV, V. B., TSIGLER, I. N., CHIRKINA, K. P., STAROSTINA, L. S., CHIRKIN, A. P., MINAEV, A. A., OREKHOVA, V. P., SEREGIN, V. F., KOLEROV, A. N., VRATSKII, A. N.: Kristallografiya **29** (1984) 963

(Received and accepted November 6, 1987)

Authors' address:

c/o Dr. P. PESHEV
Institute of General and Inorganic Chemistry
Bulgarian Academy of Sciences
Sofia 1040
Bulgaria

Book Review

H. F. Franzen

Physical Chemistry of Inorganic Crystalline Solids

Springer-Verlag, Berlin/Heidelberg/New York/London/Paris/Tokyo 1986, 158 S., 90 Abb., DM 92.–, ISBN 3-540-16580-0

Dieses sehr lesenswerte und anregende Buch resultiert aus Vorlesungen, die der Autor vor Festkörperchemikern an der Iowa-Staatsuniversität, der Arizona-Staatsuniversität und am Max-Planck-Institut für Festkörperforschung in Stuttgart vor einigen Jahren hielt. Damit ist etwa auch der spezielle Interessentenkreis umrissen. Den wesentlichen Anstoß für seine Betrachtungen und Darstellungen sieht der Autor in der großen Fülle ungewöhnlicher neuer kristalliner Materialien und den erwachsenen Fragen zum Verständnis ihrer Kristallstrukturen und kristallstrukturellen Beziehungen. Diese neuen Materialien verlangen eine weitergehende Kennzeichnung und Deutung, wofür das Hauptfeld in der Festkörperphysik und Kristallographie gesehen wird. So ist der Inhalt auch dargelegt und vermittelt kristallographische Grundlagen und Folgerungen in Anwendung auf kristallchemische und thermodynamische Probleme.

Der Inhalt ist in zehn Kapitel gegliedert. Nach einer allgemeinen Einleitung folgt eine Einführung in die Raumgitter- und Raumgruppensymmetrie. Das vierte Kapitel vermittelt eine ganze Reihe anorganischer Strukturtypen im wesentlichen aus dem Konzept der Kugelpackungen. Es folgt eine Einführung in die thermodynamische Formulierung und Behandlung heterogener Gleichgewichte, von Phasendiagrammen, Nichtstöchiometrien und Defekten. Nach einer gruppentheoretischen Einführung des reziproken Raumes und der Raumgruppendarstellung werden Aspekte von Phasenumwandlungen zweiter Ordnung auf der Basis der Landau-Theorie behandelt. Es folgt ein Kapitel über die theoretische Ableitung der Beugung an Kristallen einschließlich von Ordnungsphänomenen. Im vorletzten Kapitel werden in Anwendung der Landau-Theorie von Symmetrie- und Phasenumwandlungen einige Beispiele von Ordnungs-Unordnungsvorgängen und von geordneten Defektstrukturen diskutiert. Den Abschluß bildet eine Einführung in die Bandtheorie und die Ansätze zur Bandstrukturberechnung in Bezug auf die Konzepte der Symmetrie und des reziproken Raumes.

Am Schluß jedes Kapitels sind einige Aufgaben und Probleme formuliert, die das Verständnis überprüfen lassen. Auch hierdurch ist nochmals der beabsichtigte Lehrcharakter unterstrichen. Das Buch ist eine sehr gut gelungene Einführung in kristallographische, gruppentheoretische und thermodynamische Grundlagen in Anwendung auf kristallchemische Probleme. In seiner sachlich klaren und didaktisch knappen Form stellt es einige Ansprüche und vermittelt es viele Anregungen. H.-J. Bautsch

Cryst. Res. Technol.	23	1988	5	647–653

F. PECHAR

Institute of Geology and Geotechnics Czechoslovak Academy of Sciences
Prague, Czechoslovakia

Structure Refinement of Paranatrolite by X-ray Diffraction

Paranatrolite, $Na_2Al_2Si_3O_{10} \cdot 3\,H_2O$, monoclinic (space group Cc), but described here the pseudo-orthorhombic setting of Fd, with $a = 18.998(3)$ Å, $b = 19.018(3)$ Å, $c = 6.579(2)$ Å, $\beta = 90.37°(1)$ in analogy to scolecite. 225 parameters, including hydrogen atoms, were refined anisotropically from 1400 X-ray diffraction reflections, the final conventional R-value being 0.036. Tetrahedral arrangement of Si and Al are of O–D character as well as the Na^+ ions. Hydrogen bonding is discussed.

Introduction

Earlier studies on the crystal structure of paranatrolite, based on X-ray powder diffraction, were published by CHAO using Gandolfi camera. On the other hand, neutron diffraction studies for natrolite and scolecite have been published in the literature (PECHAR et al.; JOSWIG et al.). This paper describes the crystal structure of the Na-member of the scolecite, an other monoclin natrolite of the same class of fibrous zeolites in detail. In order to preserve the geometrical analogy to natrolite, which belongs to space group Fdd2, the pseudo-orthorhombic setting of Fd for paranatrolite was given preference over the conventional space symmetry of Cc, which, for instance, FÄLTH and HANSEN used in their X-ray analysis for scolecite. The symmetry symbol Fd also traces the diamond-like arrangement of the lateral tiepoints between the alumo-silicate fibres.

Experimental

A clear and colourless single fragment measuring $0.1 \times 0.16 \times 0.2\ mm^3$ from locality Staré Ransko, Bohemia, was prepared from a twined specimen.

Its chemical composition expresses the crystallochemical formula: $(Na_{1.65}Ca_{0.25}O_{0.05})/(Al_{2.02}Fe_{0.05}Si_{2.99})O_{10}/2.96\ H_2O$ or, ideally $Na_2Al_2Si_3O_{10} \cdot 3\,H_2O$, $D_{obs.} = 2.19$, $D_{calc} = 2.12$.

The diffraction data were collected on the automatic four-circle X-ray diffractometer SYNTEX $P_{\bar{1}}$. The wavelength of the $MoK_{\bar{\alpha}}$ radiation is 0.71073 Å. The intensity data were collected by intervals $\sin\theta/\lambda = 0.65\ Å^{-1}$. Altogether, 1400 independent reflection were measured, representing nearly 25% of the possible reflections in the $MoK_{\bar{\alpha}}$ sphere. Reflections with $I \leqq 2\sigma(I)$ were considered unobserved. Absorption was determined from transmission experiment yielding a linear absorption coefficient (f) $\mu = 2.8\ mm^{-1}$. Correction of absorption was therefore neglected. Paranatrolite crystallises pseudo-orthorhombic with 8 formula units $Na_2Al_2Si_3O_{10} \cdot 3\,H_2O$ per unit cell. The lattice parameters were determined from the adjusted settings of 25 independent reflections by least-squares method. The observed systematic extinction of reflections are in agreement with space group (Fd(Cc)).

Table 1 lists the results in comparison to previous publication.

Table 1
Comparison of lattice parameters for paranatrolite, natrolite and scolecite

Locality source	Staré Ransko Bohemia this work	St. Hilaire Quibec CHAO 1980	Natrolite Zálezly Bohemia PECHAR 1983	Scolecite Iceland JOSWIG et al. 1984
a/Å	18.998(3)	19.07(1)	18.326(5)	18.508(5)
b/Å	19.081(3)	19.13(1)	18.652(5)	18.981(5)
c/Å	6.579(2)	6.580(3)	6.601(3)	6.527(2)
β/Å	90.37(1)	—	90	90.64(1)
V/Å³	2391	2400	2256	2292

Structure refinement

Starting parameters for the refinement were taken from the natrolite structure, these atoms have numbers Si 1, 2; A 1; O1, O3, O5, O7, O9, O11, O13, H1 and H2 (PECHAR). The additional atoms result from the reduction of symmetry and are denoted with other numbers. For all calculations the XTAL package of computer programs was used (STEWART, HALL). The hydrogen positions were derived from difference Fourier maps from the natrolite refinement using the values from NMR experiment. The refinement converged to a final $R(F) = 0.036$ and $R_w(F) = 0.034$. The final po-

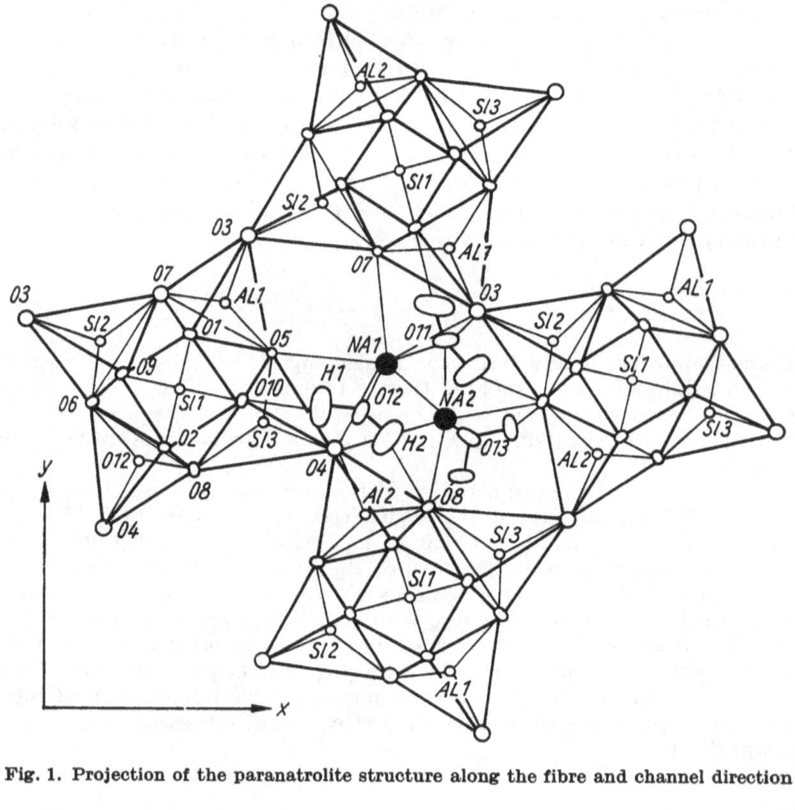

Fig. 1. Projection of the paranatrolite structure along the fibre and channel direction

Table 2a
Final atomic fractional coordinates of paranatrolite

Atom	x	y	z
Si1	0.0000(0)	0.0044(3)	0.0000(0)
Si2	−0.1043(2)	0.0443(3)	0.3697(2)
Si3	0.0887(2)	−0.0435(3)	0.3750(2)
Al1	0.0360(2)	0.0912(3)	−0.3893(2)
Al_2	−0.0385(2)	−0.0867(3)	−0.03866(2)
O1	0.0153(3)	0.0740(3)	−0.1318(3)
O2	−0.0166(3)	−0.0616(3)	−0.1443(3)
O3	0.0688(3)	0.1736(3)	−0.4251(3)
O4	0.1649(3)	−0.0801(3)	0.3895(3)
O5	0.0897(3)	0.0261(3)	0.5237(3)
O6	−0.1047(3)	−0.0204(3)	0.5303(3)
O7	−0.0509(3)	0.1045(3)	0.4585(3)
O8	0.0256(3)	−0.0943(3)	0.4569(3)
O9	−0.0705(3)	0.0178(3)	0.1524(3)
O10	0.0715(3)	−0.0162(3)	0.1386(3)
O11	0.0288(5)	0.2000(5)	0.0616(5)
O12	−0.0528(5)	−0.2041(5)	0.1633(5)
O13	−0.1887(5)	−0.0689(5)	0.1077(5)
H1	0.0254(6)	0.1551(7)	−0.0056(5)
H2	0.0603(6)	0.2283(6)	−0.0167(5)
H3	−0.0902(6)	−0.2152(6)	0.2548(5)
H4	0.2293(6)	−0.0749(6)	−0.0142(5)
H5	−0.1406(6)	−0.0644(6)	0.0853(5)
H6	−0.2019(6)	−0.1165(6)	0.1279(5)
Na1	0.2180(2)	0.0181(2)	−0.3859(3)
Na2	−0.2182(2)	−0.0182(2)	0.1140(3)

Table 2b
Anisotropic thermal parameter ($U \times 100$ Å^2) of paranatrolite defined as $T = \exp[-2\pi(U_{11}a^{*2}h^2 + U_{22}b^{*2}k^2 + U_{33}c^{*2}l^2 + 2U_{12}a^*b^*hk + 2U_{13}a^*c^*hl + 2U_{23}b^*c^*kl)]$

Atom	U_{11}	U_{22}	U_{33}	U_{12}	U_{13}	U_{23}
Si1	0.64(6)	0.68(6)	0.61(6)	0.02(5)	0.01(5)	−0.00(5)
Si2	0.49(7)	0.64(7)	0.76(7)	−0.02(5)	−0.03(5)	−0.10(5)
Si3	0.49(6)	0.64(6)	0.65(6)	−0.18(5)	0.09(5)	0.01(5)
Al1	0.58(8)	0.56(7)	0.72(7)	0.01(6)	0.05(6)	−0.04(6)
Al2	0.62(8)	0.53(8)	0.76(8)	−0.05(6)	0.08(6)	−0.03(6)
O1	1.72(5)	0.84(5)	0.79(5)	−0.06(4)	0.22(4)	0.18(4)
O2	1.97(6)	0.81(5)	1.12(5)	−0.05(4)	−0.27(4)	−0.28(4)
O3	0.54(5)	0.70(5)	1.31(5)	−0.13(4)	−0.11(4)	−0.11(4)
O4	0.68(5)	0.75(5)	1.42(5)	−0.21(4)	0.02(4)	0.06(4)
O5	1.11(5)	0.86(5)	1.30(5)	0.30(4)	−0.00(4)	−0.46(4)
O6	1.26(5)	1.10(5)	1.40(5)	0.29(4)	0.15(4)	0.51(4)
O7	0.80(5)	0.99(5)	1.55(5)	0.18(4)	−0.43(4)	−0.06(4)
O8	0.94(5)	1.27(5)	1.36(5)	0.31(4)	0.44(4)	0.08(4)
O9	0.95(5)	1.68(5)	1.22(5)	−0.20(4)	0.37(4)	0.18(4)
O10	0.80(5)	1.72(5)	0.88(5)	0.00(4)	−0.08(4)	−0.28(4)
O11	5.50(10)	2.28(7)	2.33(7)	−0.55(8)	1.60(7)	−0.99(7)
O12	2.81(8)	4.22(9)	2.08(7)	−0.69(7)	1.10(6)	−1.02(7)
O13	1.60(7)	1.38(7)	4.34(8)	0.05(5)	−0.45(6)	0.50(6)
H1	5.66(17)	2.98(14)	3.44(14)	−0.08(11)	1.18(12)	−1.08(12)
H2	7.38(22)	3.98(16)	4.88(18)	−1.56(16)	2.58(16)	−0.82(14)
H3	4.02(16)	5.36(17)	3.46(14)	−0.16(14)	1.80(12)	−0.44(12)
H4	4.64(19)	6.81(20)	4.46(16)	−1.42(16)	0.54(14)	−1.72(16)
H5	2.64(16)	5.80(19)	6.82(22)	0.12(13)	0.32(14)	0.44(16)
H6	4.74(16)	2.10(14)	4.76(16)	0.22(12)	−0.88(12)	−0.02(12)
Na1	1.02(8)	1.04(8)	1.22(8)	−0.08(6)	−0.14(6)	−0.18(6)
Na2	1.04(8)	1.02(8)	1.24(8)	−0.10(6)	−0.16(6)	−0.20(6)

Table 3a
Interatomic distances (Å), and angles (°) with their standard deviations in parentheses

Bond	distance	mean	Bond	distance	mean
Si1 – O1	1.6135(6)	1.6490	O13 – H5	0.9305(10)	
– O2	1.6071(6)		O13 – H6	0.9518(10)	
– O9	1.6976(6)				
– O10	1.6778(6)		Na1 – O3″	2.6065(5)	2.4866
			O4	2.5870(4)	
Si2 – O3	1.6096(7)	1.6307	O5	2.5099(5)	
– O6	1.6250(5)		– O7″	2.5351(5)	2.4866
– O7	1.6378(6)		– O11′	2.3864(4)	
– O9	1.6504(4)		O12	2.3521(4)	
			– O13′	2.4290(5)	
Si3 – O4	1.6096(6)		Na2 – O3	2.3801(6)	2.5377
– O5	1.6495(6)		– O4′	2.9097(5)	
– O8	1.6356(6)		– O6	3.4751(7)	
– O10	1.6705(4)		O11″	2.3777(4)	
			– O12′	2.6040(5)	
Al1 – O1	1.7762(4)	1.7303	O13	1.1189(6)	
– O3′	1.7019(7)				
– O5	1.6983(6)		O11 – O6	3.1907(5)	2.9201
– O7	1.7447(4)		O11 – O1	2.7314(5)	
			O12 – O10	2.8752(5)	
			O12 – O8	3.2019(4)	
Al2 – O2	1.7678(4)	1.7516	O13 – O2	2.7183(4)	
– O4′	1.7421(6)		O13 – O9	2.8031(5)	2.0309
– O6	1.7412(6)				
– O8	1.7554(5)		H1 – O1	1.7656(1)	2.0309
			H2 – O6	2.2504(2)	
O11 – H1	0.9670(10)	0.9531	H3 – O10	1.9324(2)	
O11 – H2	0.9587(9)		H4 – O8	2.2871(2)	
O12 – H3	0.9579(9)		H5 – O9	2.1026(1)	
O12 – H4	0.9520(10)		H6 – O2	1.8472(1)	

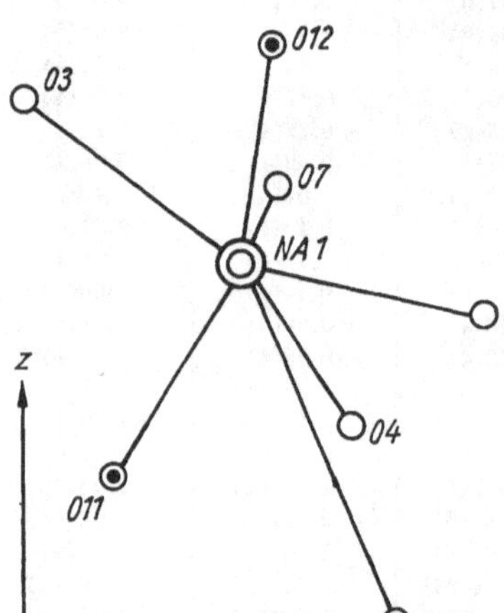

Fig. 2. Projection along *b* of the Na1 polyhedron in one of the channels

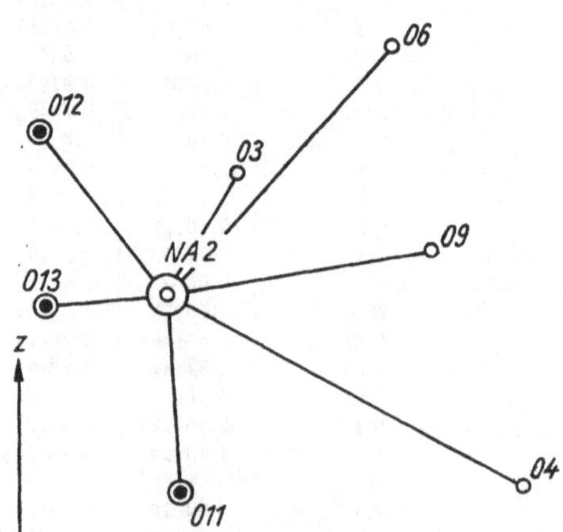

Fig. 3. Projection along *b* of the Na2 polyhedron in one of the channels

Table 3b

Bond	angle	Bond	angle
O2 – Si1 – O1	111.28(3)	O3 – Si2 – O6	113.59(4)
O2 – Si1 – O9	104.24(3)	O3 – Si2 – O7	110.97(4)
O2 – Si1 – O10	107.01(3)	O3 – Si2 – O9	108.40(4)
O1 – Si1 – O9	109.82(3)	O6 – Si2 – O7	107.84(4)
O1 – Si1 – O10	109.69(3)	O6 – Si2 – O9	109.51(4)
O9 – Si1 – O10	110.69(3)	O7 – Si2 – O9	106.27(4)
O4 – Si3 – O5	107.89(3)	O5 – Al1 – O3	114.67(2)
O4 – Si3 – O8	112.59(3)	O5 – Al1 – O1	112.97(2)
O4 – Si3 – O10	111.15(3)	O5 – Al1 – O7	127.57(5)
O5 – Si3 – O8	106.70(3)	O3 – Al1 – O1	112.79(2)
O5 – Si3 – O10	107.61(3)	O3 – Al1 – O7	101.85(5)
O8 – Si3 – O10	110.61(3)	O1 – Al1 – O7	118.35(4)
O6 – Al2 – O4	110.66(3)	H1 – O11 – H2	106.99(3)
O6 – Al2 – O2	107.99(3)	H3 – O12 – H4	106.53(3)
O6 – Al2 – O8	111.61(3)	H5 – O13 – H6	111.72(4)
O4 – Al2 – O2	107.02(3)	O2 – H6 – O13	150.85(4)
O4 – Al2 – O8	101.05(3)	O9 – H5 – O13	131.07(4)
O8 – Al2 – O2	106.24(2)	O8 – H4 – O12	118.78(5)
O12 – Na1 >O11	74.71(4)	O10 – H3 – O12	167.54(4)
O12 – Na1 – O13	50.02(4)	O6 – H2 – O11	166.58(4)
O12 – Na1 – O5	64.92(4)	O1 – H1 – O11	75.31(3)
O12 – Na1 – O7	103.44(5)	O11 – Na1 – O4	89.67(4)
O12 – Na1 – O4	92.13(4)	O11 – Na1 – O3	85.36(4)
O12 – Na1 – O3	84.35(4)	O13 – Na1 – O4	53.70(4)
O11 – Na1 – O13	66.13(4)	O13 – Na1 – O5	97.40(4)
O11 – Na1 – O5	97.40(4)	O13 – Na1 – O7	94.80(4)
O11 – Na1 – O7	88.67(4)	O13 – Na1 – O3	76.54(5)
O5 – Na1 – O7	80.46(4)	O13 – Na2 – O11	116.83(5)
O5 – Na1 – O4	91.02(5)	O13 – Na2 – O12	99.45(5)
O5 – Na1 – O3	67.01(5)	O13 – Na2 – O3	112.08(5)
O7 – Na1 – O3	86.62(5)	O13 – Na2 O6	62.87(4)
O7 – Na1 – O4	86.70(5)	O13 – Na2 – O4	90.64(4)
O4 – Na1 – O3	83.20(5)	O13 – Na2 – O9	73.93(4)
O11 – Na2 – O12	136.54(6)	O12 – Na2 – O3	83.86(4)
O11 – Na2 – O3	124.96(6)	O12 – Na2 – O4	124.85(6)
O11 – Na2 – O6	62.82(4)	O12 – Na2 – O6	77.64(4)
O11 – Na2 – O4	70.77(4)	O12 – Na2 – O9	91.74(4)
O11 – Na2 – O9	93.51(4)		
O9 – Na2 – O3	59.19(5)	O6 – Na2 – O4	88.45(4)
O9 – Na2 – O6	48.60(6)	O6 – Na2 – O3	80.97(5)

Symmetry code:

O3′: 0.0688; 0.1736; −0.4251
O7′: −0.0509; 0.1045; −0.4585
O13′: 0.3113; −0.0689; −0.3923
O3″: 0.3188; 0.0764; −0.1751
O12′: −0.3028; −0.0459; 0.4133

O4′: −0.0851; −0.1699; −0.3605
O11′: 0.2788; 0.500; 0.500; −0.1884
O7′: 0.1991; 0.1455; −0.2915
O11″: −0.2212; 0.0500; −0.1884

sitional and thermal parameters are listed in Table 2, the interatomic distances and bond angles in Table 3. Figure 1 describes projection of the fragment of structure along *b*-axis and Figures 2, 3 the coordination polyhedra of Na atoms.

Discussion

$\varrho = Si/(Si + Al) = 0.592$ ratio obtained by chemical analysis replies to partially disordered distribution of tetrahedral atoms (scolecite $\varrho = 0.599$ and natrolite 0.601). The space group of the unit cell of crystal responds to pseudoorthorhombic group Fd (or general Cc) as in the case of scolecite. But channel cation of the mineral differs from scolecite by the presence of Na^+ ions and 3 molecules H_2O per asymmetric unit of crystal. This circumstance causes the distortion of lattice parameters ***a***, ***b*** (Tab. 1) compared to natrolite and scolecite and leads to decreasing of the crystal symmetry (Fig. 1).

This deformation responds to the mean interatomic distances of T–O bonds (Tab. 3) $d(\overline{\text{Si–O}}) = 1.6403$ Å and $d(\overline{\text{Al–O}}) = 1.7410$ Å which are higher for Si–O and smaller for Al–O than values given by BAUR, $d(\overline{\text{Si–O}}) = 1.623$ Å and $d(\overline{\text{Al–O}}) = 1.752$ Å for these bonds in silicates and $d(\overline{\text{Si–O}}) = 1.617$ Å, $d(\overline{\text{Al–O}}) = 1.745$ Å than values given by FÄLTH and HANSEN for scolecite. The tetrahedric angles are adequately deformed (Tab. 3).

The mean interatomic distances O–H = 0.953 Å and angles 108.4° in water molecules respond to the usual values for crystallohydrates (HAMILTON, IBERS).

All water molecules are connected via hydrogen bonds to the framework. There is one relatively strong bond between water molecule O13 and O2 with an H ... O distance of 1.8472 Å. Its second H bond, however, has to be classified as a weak interaction despite the O ... O distance of 2.7183 Å, due to the unfavourable angle of 150.85°. In this water molecule the H–O–H angle of 111.72° deviates considerably from the angle of H_2O in vapour which is 104.5°. The other two independent water molecules are also distorted with angles 106.99° and 106.54°. The least distorted molecule O12, O11 shows the weakest hydrogen bonds to the framework (Tab. 3). No hydrogen bonds exist between the molecules of crystal water.

The coordination polyhedron Na1 (Fig. 2) consists of 4 oxygen atoms to the framework and 3 oxygen atoms to water molecules with mean interatomic distance $d(\overline{\text{Na1–O}}) = 2.4866$ Å and angle 79.52°.

Polyhedron Na2 consist of 3 oxygen atoms from water and 4 oxygen atoms from framework (Fig. 3) with mean interatomic distance $d(\overline{\text{Na2–O}}) = 2.5377$ Å and angle 88.39°. This polyhedron is considerably deformed because of the longer interatomic distances to tetrahedric oxygen and very short distance to O13.

References

BAUR, W. H.: In: Structure and Bonding in Crystals, Eds M. O. KEEFE and A. NAVROTSKY, New York 1981, 31–52

CHAO, G. Y.: Quebec Canad. Mineral. **18** (1980) 85

FÄLTH, L., HANSEN, S.: Acta Cryst. **A35** (1979) 1877

HAMILTON, W. C., IBERS, J.: Hydrogen Bonding in Solids, ed. W. A. BENJAMIN, New York 1968

JOSWIG, B., BARTL, H., FUESS, H.: Z. Krist. **166** (1984) 219
PECHAR, F., SCHÄFER, W., WILL, G.: Z. Krist. 164 (1983) 19
STEWART, J. M., HALL, S. R. (eds.): X-TAL '86, the XTAL System of Crystallographic Programs, University of Maryland, CSTRS 20742, USA 1986

(Received August 4, 1987;
accepted October 22, 1987)

Authors' address:

Institute of Geology and Geotechnics
Czechoslovak Academy of Sciences
V Holešovičkách 41
CS-18209 Prague 8
Czechoslovakia

Book Review

T. J. COUTTS, J. D. MEAKIN (eds)

Current Topics in Photovoltaics — Vol. 2

Academic Press 1987, Harcourt Brace Jovanonich Publishers, London-Orlando-San Diego-New York-Austin-Boston-Sydney-Tokyo-Toronto, xiv + 210 S., zahlreiche Abbildungen und Tabellen, Sachregister, Preis £ 38/$ 69, ISBN 0-12-153412-X.

Der zweite Band der Reihe "Current Topics in Photovoltaics" enthält 3 Artikel, wobei der Schwerpunkt auf Solarzellen aus einkristallinem Silicium liegt.

Im ersten Kapitel (80 S.) beschreiben F. A. LINDHOLM, A. NEUGROSCHEL und K. MISIOKOS die physikalischen Grundlagen, Meßmethoden und die Designtheorie von Siliciumsolarzellen. Der zweite Teil von M. A. GREEN (61 S.) behandelt spezielle Bauelementvarianten (metal-insulator-NP junction und passivated-emitter silicon solar cells). Schwerpunkte der ersten beiden Artikel sind vor allem Aspekte der theoretischen Halbleiterphysik (Einfluß der Dotierung, intrinsische Grenzen für den Wirkungsgrad, Erreichen einer hohen Leerlaufspannung usw.) und die Gestaltung (Optimierung) der Solarzellen auf der Basis von einkristallinem Si mit dem Ziel, Wirkungsgrade von mehr als 20% zu erzielen.

D. FLOOD und H. W. BRANDHORST geben im dritten Kapitel (60 S.) einen Überblick zu wichtigen Gebieten der Anwendung von Solarzellen im Weltraum. Dabei werden außer Silicium auch GaAs und InP als Materialien in Betracht gezogen, besonders unter dem Aspekt ihrer hohen Resistenz gegenüber Strahlenschäden. Neben Bauelementen auf der Grundlage von Einkristallen sind Dünnschichtvarianten und Kaskadensolarzellen einschließlich ihrer Darstellungsmethoden beschrieben.

Das Buch liegt in gewohnt guter Ausführung vor, viele Abbildungen und Tabellen sind enthalten, ein Sachwortverzeichnis erleichtert die Orientierung. Jedes Kapitel ist von führenden Spezialisten geschrieben und mit zahlreichen Literaturangaben versehen, die den neuesten Stand repräsentieren.

Die beiden ersten Beiträge des Buches sind jedoch sehr speziell gehalten und erfüllen nur teilweise die im Vorwort zum ersten Band (siehe Cryst. Res. Technol. **21**, 546 (1986)) formulierten Ansprüche für diese Serie. Wenn die Arbeiten nicht auf den Interessenkreis von relativ wenigen Spezialisten ausgerichtet werden sollen, müssen mehr Beiträge mit Übersichtscharakter aufgenommen werden, wie das im Band 1 weitgehend realisiert wurde, und wofür das dritte Kapitel im Vol. 2 ein Beispiel ist. B. SCHUMANN

Cryst. Res. Technol.	23	1988	5	655–667

V. F. SHAMRAI, D. YA. GALPEROVICH, YU. V. EFIMOV, T. V. KORZHAKOVA, A. A. BABAREKO, V. I. GRECHKOV, M. E. SAVELYEVA

A. A. Baikov Institute of Metallurgy, USSR Academy of Sciences
Moscow, USSR

T_c Anisotropy and Texture of Fine Niobium Wire and Foil

Texture of surface and inner layers of cold-deformed (> 90%) niobium wire (diameter — 0.5 mm) and foil (thickness — 40 μm) was studied immediately after deformation followed by layer-wise etching of near-the-surface layers and the annealing effect in 0.001 and 0.00001 Pa vacuum on texture was found. Axis and inner plane textures of cold-deformed niobium wire and foil, respectively, are typical of bcc lattice metals. The surface layer of fine wire subjected to considerable cold deformation possesses so-called cylindrical plane deformation texture {100}⟨011⟩ (main) and {111}⟨011⟩ (additional). Surface layer of high deformed niobium foils is characterized by the shear texture {110}⟨001⟩. Presence of delineated texture and its maintainance during annealing (with changing ratio of individual components) causes the T_c anisotropy in fine niobium materials. T_c anisotropy is caused by different orientations of magnetic fields and directions of currents used for measurements by resistive and inductive methods with respect to the directions of prefered crystallographic orientation of wire axis and foil plane. The T_c values of pure niobium wire and foil measured by the resistive method are 0.1—0.3 K higher than the standard (9.2—9.3 K).

Изучена текстура поверхностных и внутренних слоев холоднодеформированной (>90%) проволоки ниобия диаметром 0,5 мм и фольги толщиной 40 мкм непосредственно после деформации и последующего послойного стравливания приповерхностных слоев, а также влияние на текстуру отжига в вакууме 0,001 и 0,00001 Па. Холоднодеформированная ниобиевая проволока и фольга имеют текстуру оси и внутренней плоскости, соответственно, типичную для ОЦК-металлов. Поверхностный слой тонкой проволоки после значительной холодной деформации обладает так называемой цилиндрической текстурой плоской деформации {100}⟨011⟩ (основная) и {111}⟨011⟩ (дополнительная). Для поверхностного слоя сильнодеформированной фольги ниобия характерна текстура сдвига {110}⟨001⟩. Наличие выраженной текстуры и ее сохранение (при изменении соотношения ее отдельных компонентов) при отжиге вызывают анизотропию $T_к$ у ниобиевых материалов тонкого сечения. Анизотропия $T_к$ обусловлена различной ориентацией магнитных полей и направлений измерительных токов при резистивном и индуктивном методах ее измерения по отношению к направлению преимущественной кристаллографической ориентации оси проволоки и плоскости фольги. Значения $T_к$ чистой ниобиевой проволоки и фольги, измеренные резистивным методом, на 0,1—0,3 К превышают ее стандартную величину (9,2—9,3 К).

Analysis of data presented by various authors shows that the T_c values of niobium vary widely depending on material purity, methods of treatment, and often measurement procedures. For instance, according to a special reference-book of the National Bureau of Standards of the USA (ROBERTS), the T_c of pure niobium is shown to vary from 7 K to 10 K. Further increase in T_c of monocrystalline niobium (up to 12 K) can be achieved at the expense of localized superconductivity, i.e. superconductivity of twinning boundaries formed during deformation at low temperatures (BOBROV,

ZORIN; KHAIKIN). The data quoted above pertain to bulk specimens, mostly (plates and wires of rather great diameter). There are practically no systematic data on T_c of fine wire and foil where the specific surface increases highly (BUCKEL; HORZ, FROMM; SAVITSKY et al. 1981). However, it is economically and technically advisable to use fine wire and foil (BUCKEL; GALPEROVICH et al.). Besides, due to its monomorphism, niobium is an excellent material to simulate various physical and metal-scientific studies.

The present work describes experimental study of T_c, ΔT_c (using resistive and inductive methods, with accuracy to ± 0.01 K) specific resistivity ϱ (using a four-contact potentiometric method), relative resistance R_{300K}/R_{10K} and $R_{293K}/R_{4.2K}$ (in a 5 tesla magnetic field) as well as some other characteristics of long-size cold-deformed and heat-treated wire (diameter 0.1—1 mm) and foil (thickness 13—100 μm) produced from double electron-beam melting ingots in conditions described by SAVITSKY et al. 1986. The wire and foil were annealed in 0.001 and 0.00001 Pa vacuum at 1000—1100 °C (1 h) after single-pass chemical surface etching in a mixture of hydrofluoric and nitric acids (depth of layer removed was 10—15 and 6—7 μm, respectively). A 4—6 μm surface layer was etched similarly after heat treatment. Texture of wire (diameter — 0.5 mm) and foil (thickness 40 μm) was studied by analysis of data used for plotting direct and inverse pole figures of wire cross-section (taken from a wire bunch) and surface (specimen was made by winding of wire on a flat base) or foil. X-ray spectra were taken in CuKα-radiation.

The interstitial impurities content in the wire and foil used in experiments was almost similar (0.098 and 0.07 mass% O; 0.06 and 0.053 mass %C; 0.011 and 0.010 mass% N; 0.001 and 0.001 mass% H, respectively). Data of scanning electron microscopy and profilometry were used to reveal longitudinal deformation relief grooves on the surface of initial and chemically etched wire and foil (microwaveness is 3—7 and 2—4 μm, respectively; depth 2—3 μm) as well as transverse microscopes that violate the uniform microwaveness of the surface. Before and after etching the foil had class 7 roughness. The foil thickness and wire diameter decreasing, the surface microrelief parameters decrease and the number of defects increases.

Use of inductive method for measuring T_c of niobium foil (thickness 13—100 μm) shows that it decreases monotonously as the foil gets thinner. With deformation extent growing, T_c of foil decreases from 9.3 K at $\geqq$50 μm to 8.9—9.1 K at 13—20 μm (Fig. 1). Similarly, with cold deformation extent coming up to $\sim$90%, T_c of wire decreases monotonously as well; sharp decrease of T_c occurs at deformation extent of >95% (Fig. 1). The superconducting transition width ΔT_c ($\leqq$ 0.05 K in the initial state) grows markedly at smaller cold-deformation extent (0.1—0.2 K at 100—200 μm thickness, 0.3—1.1 K at 13—20 μm). ΔT_c of niobium wire in the initial state is less than 0.12 K; with cold deformation extent increasing, its ΔT_c grows as well. Thus,

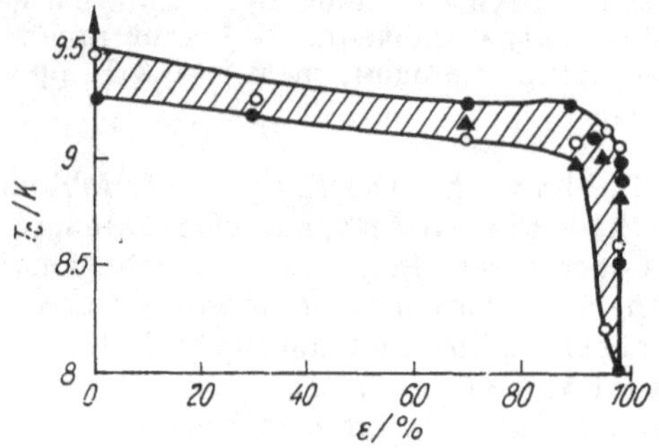

Fig. 1. Cold deformation extent vs T_c of niobium (circle-monocrystalline niobium wire; blackdot and triangle-polycrystalline electron-beam niobium wire and foil, respectively)

Table 1
T_c of niobium wire and foil measured by resistive method (K)

Material	type of specimen											
	cold-deformed with chemically etched surface				heat-treated (1100 °C, 1 h)				heat-treated with chemically reetched surface			
	$T_{beg.}$	T_{end}	ΔT_c*)	T_c*)	$T_{beg.}$	T_{end}	ΔT_c*)	T_c*)	$T_{beg.}$	T_{end}	ΔT_c	T_c
wire (diameter 0.5 mm	9.47	9.35	0.12(0.1)	9.41(9.36)	9.25	9.15	0.1(0.2)	9.2(8.8)				
foil (thickness 50 μm	9.65	9.4	0.25(0.15)	9.52(9.2)	9.31	9.24	0.07	9.27	9.42	9.29	0.13	9.355
40 μm	9.4	9.2	0.2(0.2)	9.3(9.1)	9.27	9.15	0.12(0.1)	9.21(9.0)	9.31	9.22	0.09	9.265
13—20 μm	9.5	9.3	0.2(0.4)	9.4(9.2)	9.2	9.08	0.12	9.14	9.15	9.05	0.10	9.10

*) T_c values measured by the inductive method are given in brackets

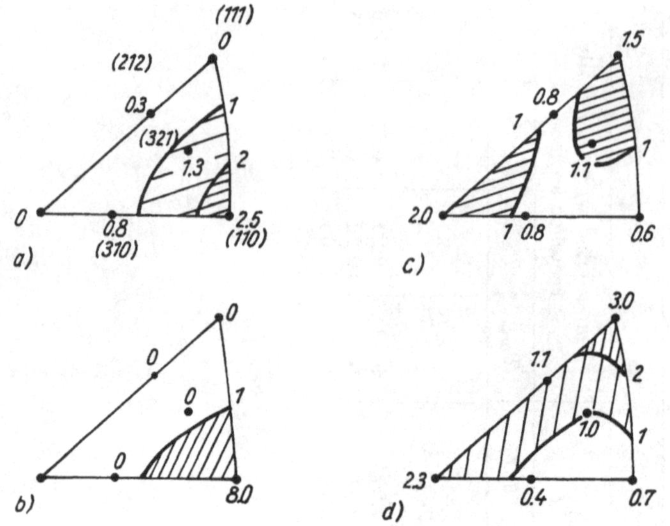

Fig. 2. Textures of cold-deformed and annealed niobium wire (diameter 0.5 mm): inverse pole figures of the axis (a, b) and surface (c, d) of cold-deformed (a, c) and annealed (b, d) niobium wire

Table 2
Texture and structure parameters of niobium wire and foil

Material and its state	wire axis	wire surface	
wire (diameter 0.5 mm)	pole density {110} in texture	texture	reflection width (400) 2 Q, °
cold-deformed	2.5	cylindrical {100}⟨011⟩ + {111}⟨011⟩	1.33
cold-deformed, after chemical surface etching	5.8	ditto	1.33
annealed	8.0	partial axial ⟨011⟩	0.50
foil (thickness 40 μm)	rolling direction	rolling plane	reflection width (400) 2 Q, °
cold-deformed surface layer (< 8 μm)	⟨001⟩ main ⟨110⟩, ⟨225⟩ additional	{110} bedding {001}, {112}, {554} additional	1.42—1.39
ditto, > 10 μm	⟨110⟩, ⟨225⟩	{001}, {112} {554}	1.33—1.30
annealed (1000 °C, 1 h) surface layer immediately after cold deformation	⟨001⟩ main ⟨110⟩, ⟨225⟩ additional	{110} bedding {001}, {554} additional	0.48
surface layer, annealing after chemical etching to a thickness of about 30 μm	⟨110⟩, ⟨225⟩	{001}, {112} {554}	0.61
annealed at > 1000 °C (1 h) surface layer, annealing after chemical etching to a thickness of about 30 μm	⟨001⟩ main ⟨110⟩, ⟨225⟩ additional	{110} bedding {001}, {554} additional	0.34

cold deformation has a noticeable effect on T_c of initially-pure monomorphous materials (niobium) only when their deformation extent is great (>90%). Distortions of the niobium lattice and growing physical defects as a result of cold deformation manifest themselves mostly by spreaded superconducting transition. Information about T_c growing as a result of cold deformation (KÖHNLEIN) is not corroborated. On the contrary, the data obtained confirm the idea expressed by GOLOVASHKIN who claims that plastic deformation of niobium brings its T_c down.

Resistive measurements show T_c of non-heat-treated wire and foil to range within 9.3—9.5 K (Tab. 1); this is more than it was recorded during inductive measurements on the same materials (9.1—9.36 K).

Cold-deformed wire and foil are strongly textured materials. In this state the wire has a cold-worked structure, the halfwidth of the last X-ray spectrum reflection (400) is 1.2—1.5°. Texture of the wire axis ⟨110⟩ is typical of bcc lattice metals (MOLLUNE). Pole texture density of axis ⟨110⟩ is 2.5—5.8 (Fig. 2, Tab. 2). The surface layer of cold-deformed wire possesses so-called cylindrical plane deformation texture

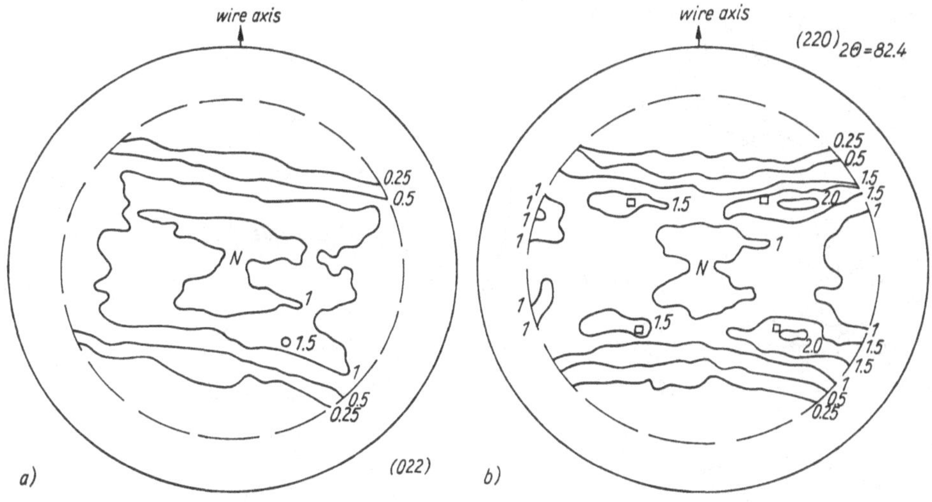

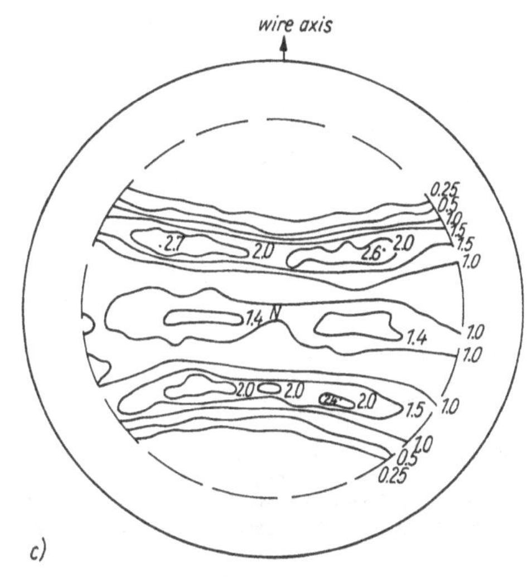

Fig. 3. Direct pole figures (220) of niobium wire surface in the initial cold-deformed state (a), after chemical etching of the surface (b) and after annealing (c)

{100}⟨011⟩(main) and {111}⟨011⟩(additional)(WASSERMAN, GREVEN). Being an independent component, the latter is to be found on the inverse pole figure of the surface only. The cylindrical texture of metal surface layer is seen most clearly by direct pole figure (220) taken from the specimen surface (Fig. 3). Designed direct pole figures plotted for cylindrical texture and partial axis texture are typical for bcc lattice metals used for type interpretation of texture revealed by the help of direct pole figures. Comparison of experimental and designed direct pole figures shows that the surface layer cylindrical texture axial blurring is $\pm 20°$ (around the direction of drawing). Surface etching by 7–10 μm somewhat improves the cylindrical surface texture (Fig. 3b). The layer of cylindrical texture of plane deformation is at least deeper than 10 μm.

Texture of cold-deformed foil (40 μm) is not uniform also. In surface layers having a depth of <8 μm it is identical with the shear texture {110}⟨001⟩ (Fig. 4), which

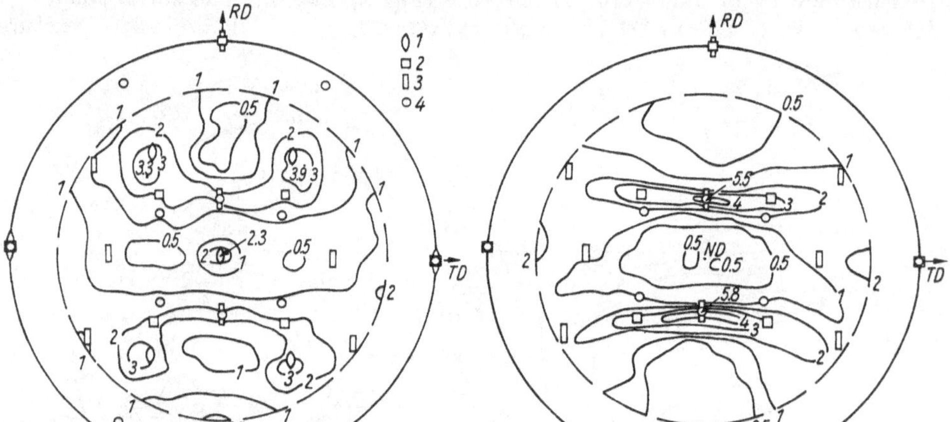

Fig. 4. Direct pole figures (220) of cold-deformed foil (thickness 40 μm): a) surface layer; b) after electrochemical etching with removal of ~ 4 μm layer from each side of foil; c) after annealing of non-etched foil; d) after annealing of etched foil. Orientations: 1 – {110}⟨001⟩, 2 – {001}⟨1$\bar{1}$0⟩, 3 – {112}⟨1$\bar{1}$0⟩, 4 – {554} ⟨22$\bar{5}$⟩. RD – rolling direction, TD – transverse direction

is characterized by localized niobium slip planes {110} in the rolling plane. As the surface layers are being removed, the texture component of plane deformation in bcc lattice metal is increased and that of shear texture weakens (Fig. 4). The plane deformation texture contains a partial axis component {001}—{112}⟨1$\bar{1}$0⟩ and also discrete orientation {554}⟨225⟩ which is close to {111}⟨112⟩ and differs from the latter by turning through 6° around the transverse direction. As the surface layers of metal are being removed, the rate of changes in fractions belonging to various foil texture components differs (Fig. 5). As the specimen gets thinner, components of texture (partial axis) {001}⟨110⟩ and {112}⟨110⟩ increase slightly, at first, then their fractions are saturated and, finally, display a tendency to go down (Fig. 5). When metal layers possessing shear texture are being removed, the plane deformation texture close to octahedral {111}⟨112⟩ one grows most intensively. On inverse pole figures of normal to rolling plane (Fig. 6) this texture component is described by plane {111} due to limited number of intensity counting points.

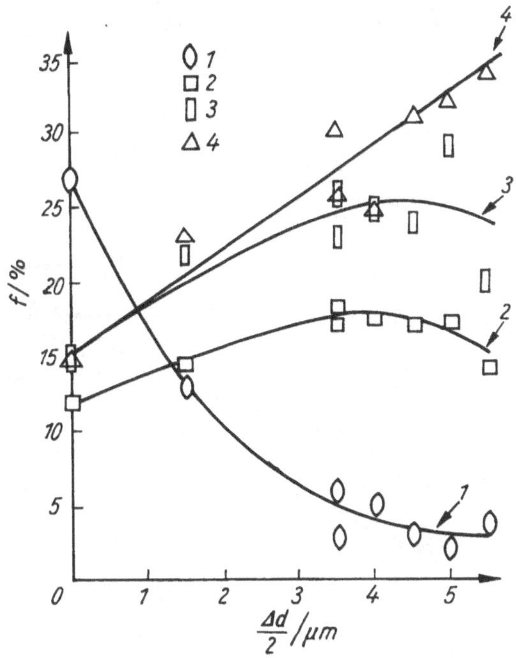

Fig. 5. Texture changes along the foil cross-section revealed by surface etching during removal of surface layer $\Delta d/2$ (μm) per side. Symbols: d – foil thickness after etching, f – fraction of components of shear texture (1) and plane deformation (2–4) according to data of analysis of inverse pole figures of normal to the rolling plane

1 – {110}, 2 – {001}, 3 – {112}, 4 – {111}

We obtained different T_c values of the same strongly-deformed materials, that are similar in cold deformation extent, composition uniformity level, impurity content and all effects of external factors in the process of production. With resistive and inductive methods of measurement used, it can be also accounted for by anisotropy of their crystalline structure.

T_c anisotropy by various crystallographic directions of hexagonal lattice in single crystals of superconducting compounds, due to the structure anisometry and corresponding anisotropy of properties in the initial state, is proved by THOLENCE et al. Anisotropy of crystalline structure of niobium wire and foil which occurs during cold deformation and is marked by the difference of reticular density in various crystallographic directions can deform the Fermi surface, increase the anisotropy of free path of electrons and as a sequence change of electric and superconducting properties can

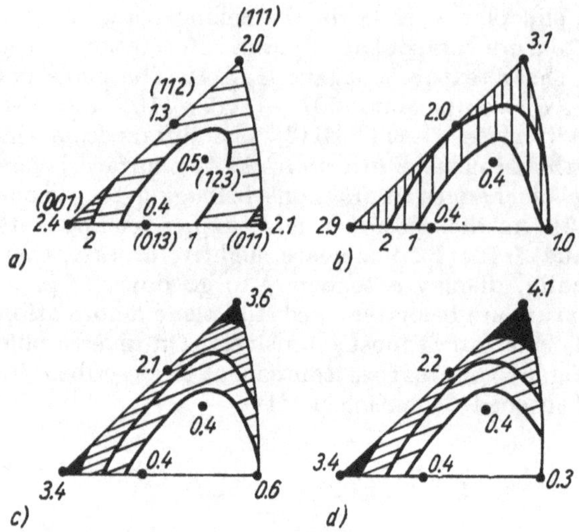

Fig. 6. Inverse pole figures of normal to the rolling plane of foil (thickness 40 μm in the initial state (a) and after chemical surface etching for 10 s (b), 15 sec (c) and 20 s (d)

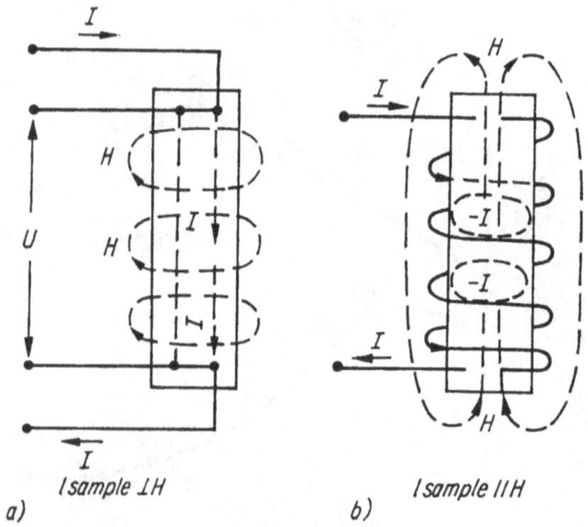

Fig. 7. Resistive (a) and inductive (b) methods of T_c measurement

be created. H_{c2} and J_c anisotropy of single-crystal and cold-deformed materials was established long ago (BUCKEL; SAVITSKY et al. 1981). Different T_c values of the same specimens on inductive and resistive measurements can increase from strong anisotropy of their structure. The establishment of differences in T_c mentioned above can be accounted for by specific features of measurements described here. With the resistive method used (Fig. 7) the current flows along the specimen length, i.e. in the direction of the preferred crystallographic orientation of strongly-textured cold-deformed material. The current used for measurements is parallel to specimen texture axis (in wire, I || ⟨110⟩) and the resultant magnetic field is perpendicular to it (H ⊥ ⟨110⟩). In case of inductive method (Fig. 7b) the material is affected by the magnetic field directed along the specimen (H || ⟨110⟩, in wire). This magnetic field excites Foucault currents in the specimen that is in the superconducting state (i.e. currents flowing perpendicular to the specimen axis — I ⊥ ⟨110⟩). It is the orthogonality of directions of currents flowing through specimens that helped to reveal the T_c anisotropy of strongly-deformed niobium wire and foil investigated.

Annealing intensifies the axis texture of cold-deformed niobium wire to $P_{\langle 110 \rangle} = 8$ (Fig. 2). Following annealing the halfwidth of reflection (400) falls down to 0.5° (scale 2Q). Reflections in the spectrum of annealed wire specimens are partly split into $K_{\alpha_1}K_{\alpha_2}$ — doublet, beginning with reflection (321). During relatively low-temperature annealing (< 900 °C) cylindrical texture of surface layer of cold-deformed wire is maintained completely or partly. High-temperature annealing (1000—1100 °C, 1 h) causes the surface layer cylindrical texture to disintegrate and pass into partial axis one which is a characteristic of the inner part of the wire (Fig. 2, 3, Tab. 2). This process can be presented as absorption of cylindrically structured surface layer grains by metal grains growing inside the wire (Fig. 8). In such case the ⟨110⟩ axis texture of the wire is almost three times as much.

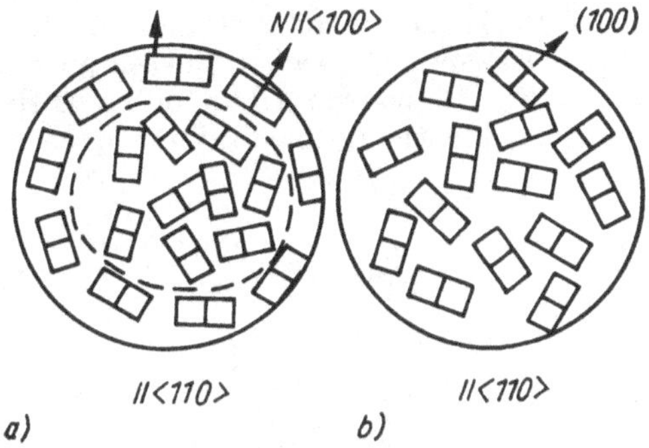

Fig. 8. Orientation of crystallite lattices in cross-section of cold-deformed (a) and annealed (b) niobium wire. Direction of drawing (wire axis) is parallel to ⟨110⟩

Texturing taking place during recrystallization of foil differs from that of wire. Following annealing at 1000 °C for 1 h the foil surface texture is maintained (Fig. 9, 4c). Thus, during such annealing regime surface shear texture grains are not absorbed by those of inner layers of foil having plane deformation texture. The portion of the latter in the surface layer becomes even smaller and the ratio of shear texture fraction to the sum of plane deformation texture ones ($f_{\text{shear}}/f_{\text{plane deformation}}$) increases from the initial value of 0.64 in the deformed foil to 0.82 following annealing. Hence a conclusion can be drawn that in these conditions the rate of growth of shear-textured grains is higher than that of plane deformation-textured ones.

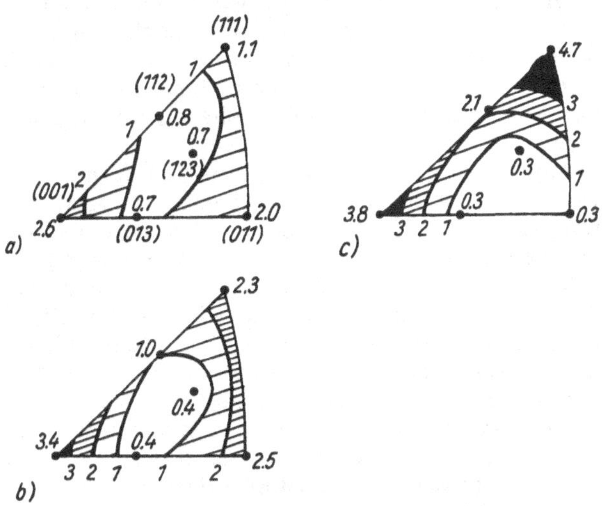

Fig. 9. Inverse pole figures of normal to rolling plane of cold-deformed foil (40 μm) after annealing at 1000 °C (1 h): a — no surface etching before annealing, b — surface etched chemically to a thickness of ~ 30 μm before annealing

In case of a 4–5 μm surface layer removed by etching prior to recrystallization so that the new surface texture contains almost no {110}⟨001⟩ orientation ($f_{shear}/f_{plane\ deformation} \sim 0.04$) the plane deformation texture is maintained after annealing by the regime mentioned, here (Fig. 9, 4d). So-called "in-situ" recrystallization takes place and recrystallized foil differs from the deformed one in the structural state only. This is revealed during X-ray investigation after the decrease of reflection widths (Tab. 2). Incomplete removal of shear-textured metal layer promotes its formation by annealing due to the predominated growth of {110}⟨001⟩ grains.

After annealing at 700–1200 °C for 1–5 h T_c of niobium foil ranges widely from 7 to 9.3 K, depending on conditions of material pretreatment and annealing regimes. T_c of fine foil (13–20 μm) annealed after deformation goes down gradually as the temperature and time of annealing in 0.001 Pa vacuum are increased (Fig. 10, curve 1). Annealing after etching of surface layer of cold-deformed fine foil causes T_c to decrease much less when the temperature and time of annealing in 0.001 Pa vacuum go up (Fig. 10, curve 2). The foil being relatively thick (e.g. 40–50 μm), T_c variations following identical annealing are even less than shown above, provided the surface

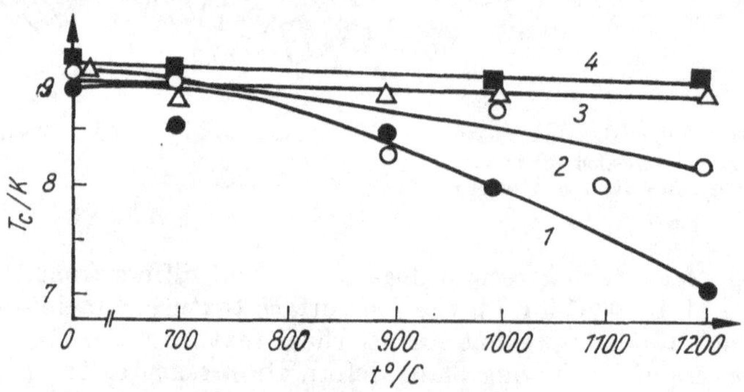

Fig. 10. Effect of temperature (annealing for 1 h) on T_c of cold-deformed niobium wire (thickness 13–20 μm). 1 no surface etching before annealing; 2 – 7–10 μm layer etched before annealing; 3 – cold-deformed foil (thickness 40 μm (3) and 50 μm (4) with surface layer etched before annealing

layer was etched (Fig. 10, curves 3 and 4, respectively). However in this case there exists a tendency for T_c to decrease slightly after annealing as well. Pre-etching of the surface-textured layer prevents development of the annealing texture with increase in the shear texture component, hence, data on the effect etching before annealing has on T_c make it possible to believe that the latter depends on the material texture, too.

According to the activation analysis data, in the process of annealing in 0.001 Pa vacuum niobium foil is penetrated by oxygen and other interstitial impurities from the residual furnace atmosphere. Absorption of oxygen and resultant decrease in T_c are corroborated by the fact that this effect gets stronger when annealing time in 0.001 Pa vacuum at high temperatures becomes longer. An equal amount of impurities can be assumed to penetrate specimens of various thickness over the same time and temperature. Fine foil has a smaller apparent cross-section than a thicker one, thus, it experiences greater structural distortions and chemical changes caused by interstitial impurities. In fact, under such conditions the finest foil can get "rancid"-through, so that no base material is left in it whatsoever.

The assumption expressed is corroborated by measurements of relative resistance (Tab. 3). Relative foil resistance decreases as it gets thinner. Thus, it becomes clear

Table 3
Relative resistance (R_{300K}/R_{10K}) of niobium wire and foil

Material	type of specimen		
	cold-deformed with chemically etched surface	heat-treated (1100 °C, 1 h)	heat-treated with chemically reetched surface
wire (diameter 0.5 mm)	30.2(30.8)*)	46.1(50.1)*)	
foil:			
50 μm	23.2	20.8	20.7
40 μm	17.9	15.7	16.6
20 μm	19.8	13.5	13.8
13 μm		6.5	

*) Values of $R_{293K}/R_{4.2K}$ measured in a 5 kOe magnetic field are given in brackets

why heat-treated foil has smaller relative resistance than the original, cold-hardened one: due to penetrating impurities the apparent cross-section of pure metal is reduced and the cross-section area occupied by the alloy as such increases. Oxygen does not penetrate the entire material depth, but is concentrated near the surface. With smaller cross-section of material, the effect of this oxygen-rich layer is more pronounced. Relative resistance of 20 μm foil is nealy 30% less as compared to the 50 μm one and that of 13 μm foil is 70% less. Relative resistance of heat-treated 13 μm foil makes it possible to assume that the oxygen-contaminated layer occupies almost its entire volume.

Heat-treated wire (diameter 0.5 mm) has a greater relative resistance than the original deformed one (Tab. 3). The wire diameter is an order of magnitude greater than the thickness of foil discussed above. The volume of oxygen-rich layer is low as compared with that of wire, hence, its influence on electrical resistance is negligible. On the other hand, heat treatment at 1000—1100 °C results in recrystallization of the entire wire volume and growth of grains; due to this, the length of grain boundaries becomes smaller, electrical resistance goes down and the value of relative resistance increases. The influence of this mechanism on electrical resistance is greater than the negative effect of penetrating oxygen.

Removal of oxygen-contaminated layer should have favourable impact on resistivity and relative resistance of fine materials. After chemical etching of heat-treated foil with removal of a 3—5 μm layer from each side of 50 μm specimen the value of relative resistance remains practically unchanged; in case of 40 μm specimen it goes up by 5% and with 20 μm specimen it is 2% higher than it used to be. Such variations of the relative resistance of foils having different thicknesses also serve to prove the considerable impact of the near-the-surface layer on fine material properties.

The effects caused by oxygen penetration practically disappear during annealing in 0.00001 Pa vacuum and more. This is in agreement with the data of SCHULZE and KREHL pertaining to prevention of oxidation and improvement of niobium properties during high-vacuum annealing.

Summary

1. Resistive and inductive methods used for measurements of T_c of greatly-textured cold-deformed niobium materials (wire, foil) showed presence of anisotropy of this

superconducting parameter as well. T_c anisotropy is due to different orientations of magnetic fields and directions of currents used for measurements with respect to preferred crystallographic orientation of wire axis and foil plane. T_c values of pure niobium wire and foil measured by the resistive method are 0.1–0.3 K higher than the standard (9.2–9.3 K).

2. Cold-deformed niobium wire has an axis texture ⟨110⟩, and foil plane deformation texture {001} + {112}⟨1$\bar{1}$0⟩ + {554}⟨22$\bar{5}$⟩ (in the rolling plane), which is typical of bcc lattice metals. However, after great cold deformation (>90%) the wire surface layer possesses so-called cylindrical plane deformation texture {100}⟨011⟩ (main) and {111}⟨011⟩ (additional). Surface layer of greatly-deformed niobium foil is characterized by shear texture {110}⟨001⟩.

3. Annealing intensifies the axial texture of niobium wire. Cylindrical surface-layer texture of cold-deformed wire is maintained completely or partly at relatively low-temperature annealing (<900 °C). At high-temperature annealing it disintegrates and passes into partial axial texture which is typical of the inner part of the wire. In case of annealing at 1000 °C for 1 h the foil surface layer texture becomes more pronounced due to pre-eminent growth of {110}⟨001⟩ grains in the matrix with plane deformation texture. With the surface layer removed by foil etching prior to annealing, so-called "in-situ" recrystallization takes place at the heating regime described above and the metal plane deformation texture is maintained.

4. Heat treatment of fine niobium materials in 0.001 Pa vacuum causes the relative resistance to decrease. Similar heat treatment of bulk niobium materials makes this parameter grow. This is accounted for by:

a) penetration of interstitial impurities, particularly oxygen, throughout the entire foil depth as a result of annealing in 0.001 Pa vacuum at high temperature (1000 to 1100 °C);
b) well-developed foil surface, and
c) small foil thickness.

It is annealing in vacuum (not worse than 0.00001 Pa) only that prevents fine niobium materials from getting contaminated by oxygen and other interstitial impurities.

References

BOBROV, V. S., ZORIN, S. N.: Abstracts of Reports delivered at the 23nd All-Union Conference on Physics of Low Temperatures; Part I, Tallin 1984

BUCKEL, W.: Supraleitung, Grundlagen und Anwendung, 2. Auflage, Weinheim 1977

GALPEROVICH, D. Ya., GANIN, S. A., GRECHKOV, V. I. et al.: Elektronnaya tekhnika (Electronic Technology) (in Russian) Ser. Electronics, **5** (1985)

GOLOVASHKHIN, A. I.: Uspekhi fiz. nauk (Advances in Physical Sciences) (in Russian) **148** (1986) 2

HÖRZ, G., FROMM, E.: Vacuum **33** (1983)

KHAIKIN, M. S.: Letters to ZETF (J. Exp. Theor. Phys.) (in Russian) **33** (1981)

KÖHNLEIN, D. Z.: Phys. **208** (1968)

MOLLUNE, W. F. et al. (ed.): Powder diffraction file search manuelled, Internat. Centre Diffraction Date, USA, Park Lane 1977

ROBERTS, B. W.: J. Phys. Chem. Reference Data **5** (1976) 3

SAVITSKY, E. M., EFIMOV, Yu. V., GALPEROVICH, D. Ya. et al.: Cryst. Res. Technol. **21** (1986) 1

SAVITSKY, E. M., EFIMOV, Yu. V., KRUZHLYAK, Ya. et al.: Fiziko-khimicheskie osnovy polucheniya sverkhprovodyaschikh materialov (Physico-chemical procedures for production of super-conducting materials) (in Russian) Moscow 1981

SCHULZE, K., KREHL, M.: Nucl. Inst. and Methods A2366 (1985)
THOLENCE, J. L., FLOUQUET, J., FRANSE, J. J. M., MENOVSICY, A.: J. Magn. Mater. **54–57** (1986)
WASSERMANN, G., GREVEN, I.: Tekstura metallicheskikh materialov (Texture of Metallic Materials) (in Russian) Moscow 1969

(Received and accepted October 22, 1987)

Authors' address:

c/o Yu. V. EFIMOV
A. A. Baikov Institute of Metallurgy
USSR Academy of Sciences
Leninskii prospekt 49
Moscow 117911, USSR

Cryst. Res. Technol.	23	1988	5	668

Book Review

O. V. MAZURIN, M. V. STRELTSINA, T. P. SHVAIKO-SHVAIKOVSKAYA (eds)

Handbook of Glass Data — Physical Science Data 15. Part C: Ternary Silicate Glasses

Elsevier Sci. Publ. B. V. Amsterdam/Oxford/New York/Tokyo 1987, 1110 S., US $ 307.25/ Dfl. 630.00, ISBN 0444-42889-5.

Vom Standard-Werk "Handbook of glass data" ist nun der Band C erschienen, der die entsprechenden Daten von ternären Silicat-Glassystemen enthält. Im vorausgegangenen Band A waren die Daten von Kieselglas und binären Silicat-Gläsern und im Band B solche von Einkomponenten- und von binären Nicht-Silicatgläsern enthalten. Folgende Glassysteme sind im vorliegenden Band C berücksichtigt worden:

$R_2O-R_2'O-SiO_2$	$RO-R'O-SiO_2$	$R_2O_3-RO_2-SiO_2$
$R_2O-RO-SiO_2$	$RO-R_2O_3-SiO_2$	$R_2O_3-R_2O_5-SiO_2$
$R_2O-R_2O_3-SiO_2$	$RO-RO_2-SiO_2$	$R_2O_3-RO_3-SiO_2$
$R_2O-RO_2-SiO_2$	$RO-R_2O_5-SiO_2$	$RO_2-R'O_2-SiO_2$
$R_2O-R_2O_5-SiO_2$	$RO-RO_3-SiO_2$	$RO_2-R_2O_5-SiO_2$
$R_2O-RO_3-SiO_2$	$R_2O_3-R_2'O_3-SiO_2$	$R_2O_5-R_2'O_5-SiO_2$
		$R_2O_5-RO_3-SiO_2$

Folgende Eigenschaften sind in Tabellen, graphischen Darstellungen und Kurven zusammengestellt:

- Glasbildungsbereiche
- Kristallisation
- Dichte
- thermische Ausdehnung und andere thermische Eigenschaften
- optische Eigenschaften
- Viskosität
- Diffusion
- Ionendiffusion
- Verdampfung
- elastische Eigenschaften und innere Reibung
- Festigkeit
- Oberflächenspannung
- chemische Beständigkeit
- elektrische Eigenschaften
- Gaslöslichkeit und Gasdurchlässigkeit
- magnetische Eigenschaften

Allein diese Aufzählung zeigt, wie umfassend die Autoren solche in der Literatur bis zum Jahre 1985 veröffentlichte Glaseigenschaften zusammengetragen und übersichtlich zusammengestellt haben. Das Werk enthält am Ende ein Literatur-, ein Autoren- und ein Glassystemverzeichnis und ist in sich geschlossen und wie die Teile A und B getrennt zu verwenden.

Man muß den Autoren dankbar sein, daß sie sich einer solch extrem umfangreichen Literaturarbeit unterzogen haben. Damit ist ein annähernd gleichwertiges Standardwerk zu den "Phase diagramms", welches von der American Ceramic Society herausgegeben wurde, entstanden und jeder Glasfachmann wird gerne zu ihm greifen.

Es sind 1711 Arbeiten ausgewertet und als Literaturstellen angegeben worden. Allerdings sind die Literaturzitate im Text nur mit Namensangaben und ohne Nummerierung etwas problematisch, so daß die konkrete Literaturstelle im Literaturtziatverzeichnis erst gesucht werden muß.

Im Vorwort wird auf die Problematik der Auswertung einer enorm hohen Anzahl von Arbeiten aufmerksam gemacht und darauf verwiesen, daß von den Autoren nur die besten Arbeiten zur Aufnahme in das Buch ausgewählt wurden. Damit erhält das Werk natürlich eine subjektive Note. Von den 1711 Literaturangaben stammen über 40% von sowjetischen Autoren. Da vielen Benutzern des Werkes diese Arbeiten zum Teil unbekannt oder nicht zugänglich waren, braucht das kein Nachteil zu sein.

W. VOGEL

Cryst. Res. Technol.	23	1988	5	669–675

S. P. S. ARYA, O. N. SRIVASTAVA

Thin Film and Electron Microscope Laboratory, Department of Physics
Banaras Hindu University

Investigation on Synthesis, Characterization, and Electronic Behaviour of Zinc Oxide Thin Films in Relation to Fabrication of Oxygen Gas Sensor

The zinc oxide thin films were prepared by spray pyrolysis method. The films were characterized by means of X-ray diffraction and scanning electron microscopy techniques. The annealing treatment at 300 ± 30 °C for 15 and 30 hours improved the crystallinity of the film and also led to a homogeneous film surface. The O_2 gas sensing properties of the zinc oxide thin film were studied by investigating the variation of the electrical resistance of the film as a function of O_2 partial pressure at various temperatures. The gas sensing property was found to depend on the film characteristics in regard to its state of crystallinity and surface conditions. The gas sensing property was also dependent on the temperature at which the sensing was done. The optimum oxygen gas sensivity was obtained when the temperature of the zinc oxide sensor was maintained at 400 °C. The study reveals that sprayed zinc oxide thin films may be used as a low cost solid state oxygen gas sensor.

1. Introduction

Recently, there has been a significant spurt in studies relating to oxide semiconductors with reference to their possible use for sensing specific gases. Oxide semiconductor materials fulfil a considerable number of the requirements for gas sensors such as high sensivity, long term stability and fast response. However, some difficulties remain in regard to selectivity and high cost. The oxide solid state gas sensors are based on the fact that the adsorption and desorption of gases cause a notable often profound change in the electrical resistance of the semiconductor. The use of ZnO, SnO_2 semiconductors for detecting various gases such as O_2, H_2, CO, etc. have been investigated (ADVANI, JORDAN; ARYA et al.; CHANG; HEILAND; MORRISON; NITTA, HARADOME; AYABI; SEIYAMA et al.; WINDISCHMANN, MARK). Reported results (HEILAND) indicate that most of the resistance changes caused by gas adsorption is related to the electron movement instead of ion movement, i.e. the chemisorbed gases lead to generation of bonding of conduction electrons. Since the change of resistance takes place in the surface, a high area to volume ratio is desirable. This requirement clearly favours the use of oxide semiconductors as a gas sensor in the form of thin films.

ZnO exhibits promising surface properties (HEILAND) and has often been used as a catalyst in chemical technology. Apart from these uses, the ZnO thin films have several other applications, also such as a surface acoustic wave r.f. filter, transparent and conducting electrode for various applications and heat mirrors etc. The properties of the film depend on the method of preparation, and there are various techniques such as sputtering, thermal oxidation, and spray pyrolysis by which ZnO thin films can be prepared. Spray pyrolysis is possibly the easiest and economically the most viable method for the preparation of an inexpensive film. A lot of work (TANSLEY, NEELY; WEBB et al.; YEN) has been done on sputtered ZnO thin films for various applications. But the investigation of sprayed ZnO thin films for gas sensor application does not seem to have been investigated so far. The present paper reports the results of our work with ZnO films prepared through spray pyrolysis and employed as gas (oxygen) sensor.

2. Experimental

The zinc oxide thin films were prepared by spraying an alcoholic solution containing 0.7 M zinc acetate on substrates at the temperature 375 °C using fine spray nozzel. The substrates used were glass slides, cut into one inch squares. The temperature of the substrate was continuously monitored by Chromel-Alumel thermocouple fixed on the top surface of the substrate, and during the spray it could be controlled within ± 5 °C with the help of temperature controller. Oxygen was used as the carrier gas, and the spray rate was about 20 $cm^3\,min^{-1}$.

The as-grown films were annealed at temperatures 300 ± 30 °C in air for time periods of (a) 15 h and (b) 30 h. The thickness of the films was measured with a Talystep instrument. All films, on which experimental data are reported here were about 5000 Å thick. The zinc oxide coated glass slide was mounted on the metal block attached with a heater. The desired temperature of the films was achieved by controlling electric power supplied to the heater, and temperature of the film was determined with a fine thermocouple attached to the film. For gas sensitivity measurements, the conventional vacuum system was used as an environment test chamber by controlling the gas leak with the pumping rates.

For electrical connections, silver paste was painted on the two corners of the sensor element. A protection resistor was connected in the series to the sensor, and d.c. voltage 4.5 v was applied to the circuit. The electrical characteristics of the sensor material were observed by recording the change of the d.c. voltage across the protection resistor.

3. Results and discussion

All films were characterized structurally through X-ray diffraction (XRD) technique Figure 1 shows the X-ray diffraction patterns of the as-prepared and annealed films. The observed peaks agreed with the known structure of ZnO (hexagonal, $a = 3.249$ Å, $c = 5.20$ Å). The number of peaks and intensity increase with annealing. It seems that the crystallinity of the film improves as a result of annealing. Also XRD data suggest that the films grown by spray pyrolysis have a dominant (00.1) orientation. This fact was evidenced by the appearance of strong 002 diffraction line for as-grown as well as annealed films.

Figure 2 shows the scanning electron micrographs of the as-deposited and annealed films. As can be seen from the figures, whereas the surface of the as-deposited films is inhomogeneous, on annealing the surface becomes rather uniform. This characteristic is brought out by the disappearance of the two different types of the surface matrix in the annealed films. It has been noted that the surface homogeneity is obtained after a specific annealing time period. As for example for the ZnO films under investigation, it was noticed that the film homogeneity remained nearly invariant after a time period of about 15 h.

The sheet resistance of the films used for sensor measured at room temperature were around 10 and 100 $M\Omega\,cm^{-2}$ for as-deposited and annealed films, respectively.

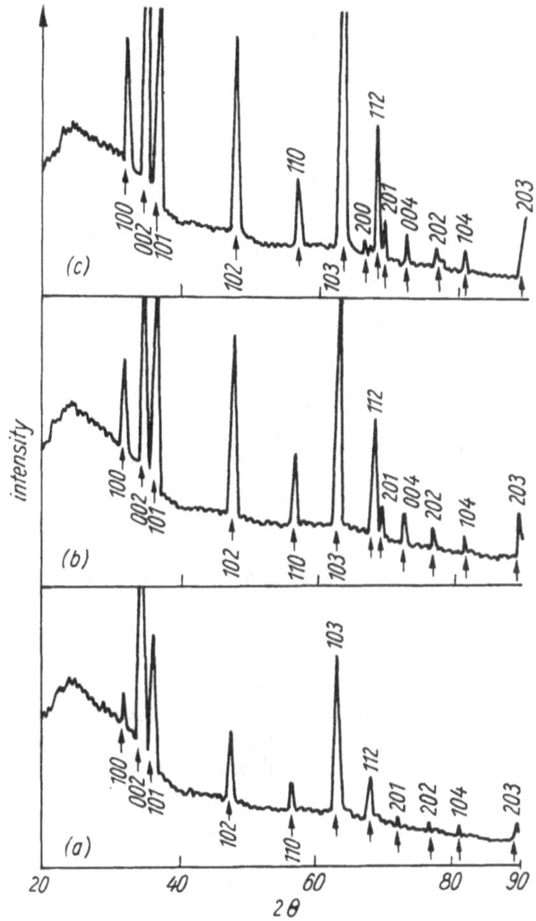

Fig. 1. X-ray diffraction patterns of ZnO thin films. (a) As-deposited films. Annealed at temperature (T) 300 ± 30 °C for time period (b) 15h, c) 30 h

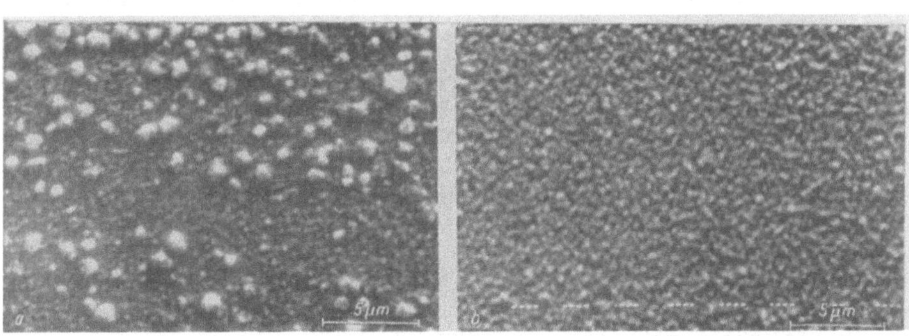

Fig. 2. Scanning electron micrograph of ZnO thin films (a) as-deposited, (b) annealed at temperature 300 ± 30 °C for time period 15 h

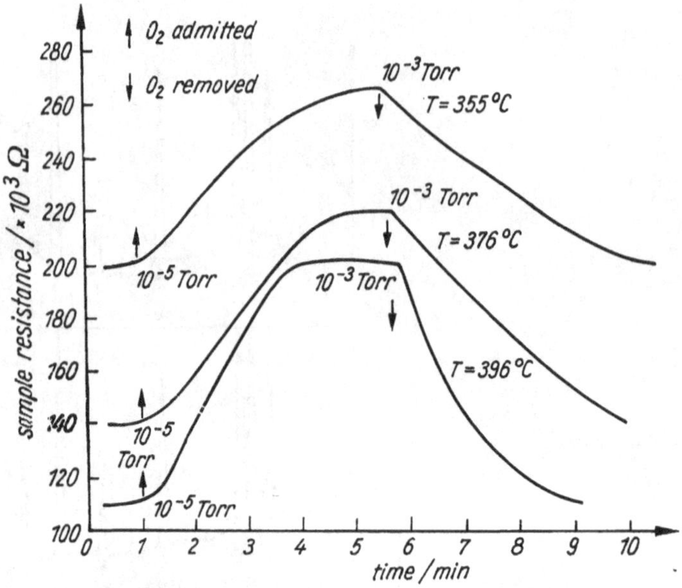

Fig. 3. Variation of resistance of ZnO thin film corresponding to O_2 partial pressure at various temperatures

When the sample was heated inside vacuum, the resistance decreased. For oxygen sensitivity measurements the sample was kept at desired temperature under a vacuum of 10^{-5} Torr. The change in the electrical resistance was measured with changing partial pressure of O_2. The results shown in Figure 3 are the variations in the resisance of the sample at temperatures 355, 376, and 396 °C, respectively when the environment changes from 10^{-5} to 10^{-3} Torr O_2 partial pressure and vice-versa. The resistance of the sample increases after O_2 introduced and becomes constant with keeping constant the partial pressure of O_2. As the O_2 pressure released from 10^{-3} to 10^{-5} Torr, the resistance decreases and recovers its original value. Results also show that the extent of the resistance variation increases as the temperature increases, the higher the temperature, the faster the resistance variation.

At temperature lower than 350 °C, the variation in the resistance on exposing to O_2 gas was very small whereas at temperatures higher than 400 °C the gas sensing characteristics as monitored by the variation of the electrical resistance was found to be irreversible. The gas sensing characteristics were investigated upto a temperature of about 400 °C which was the optimum temperature for reversible resistance variations. Figure 4 shows the variation of resistance when the environment changes from 10^{-5} to 10^{-1} Torr O_2 partial pressure and vice-versa at 396 °C for the as-deposited and annealed films. The resistance ratio corresponds to the resistance under 10^{-1} Torr O_2 partial pressure, and 10^{-5} Torr vacuum is 7, 20, and 10 for the as-deposited and annealed 15 and 30 h duration. This shows that there is a large variation in resistance on exposure to O_2 gas for the film annealed at 15 h in comparison with other films. Also the rate of resistance variation for this film was faster than others. It was also found that for the films annealed for 30 h, the electrical resistance variation on O_2 exposure was not easily reproducible. Similar was the case with the as-grown films. In sharp contrast to this, for the film annealed for 15 h the variation in resistance was quite reproducible.

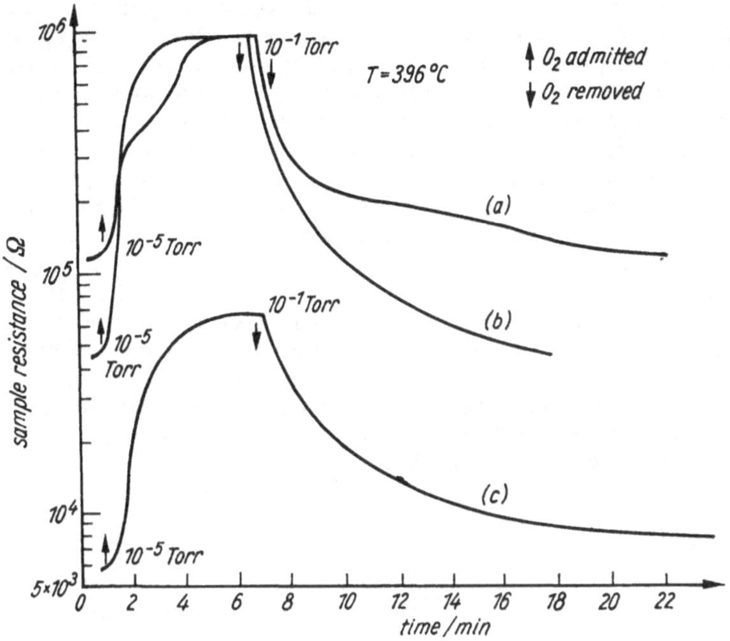

Fig. 4. Variation of resistance of ZnO thin film corresponding to O_2 partial pressure at temperature 396 °C. (a) As-deposited film, annealed at temperature 300 ± 30 °C for time period (b) 15h, (c) 30 h

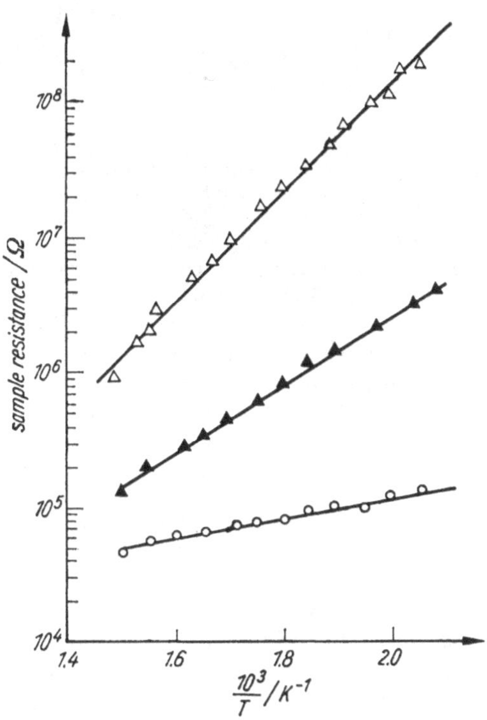

Fig. 5. Variation of resistance of 15 h annealed ZnO thin film with the reciprocal of the temperature at various O_2 partial pressure: △ 10^{-1} Torr, ▲ 10^{-3} Torr, ○ 10^{-5} Torr

Resistance of the film annealed for 15 h and exposed to various temperatures was also measured. First the temperature of the sample was raised to approximately 400 °C at a given O_2 pressure and then variation of the resistance with temperature was measured. The results are shown in Figure 5. The resistance at each temperature is a strong function of oxygen partial pressure, and there exists a linear relationship between the logarithm of the sample resistance and the reciprocal of the sample temperature. The slope of the straight line changes as the partial pressure of O_2 changes. Taking activation energy to be governed by an Arrhenius equation, the calculated activation energies from the slopes came out to be 0.15, 0.45, and 0.85 ev corresponding to O_2 partial pressure of 10^{-2}, 10^{-3}, and 10^{-1} Torr, respectively.

The resistance of the film annealed for 15 h varies linearly with changing O_2 partial pressure at a temperature of 396 °C as shown in Figure 6. These results reveal that this film can be used as an O_2 gas partial pressure sensor. The adsorption and desorp. tion times for this film are ~ 2 and 10 min when O_2 gas is cycled from 10^{-5} to 10^{-1} Torr. The resistance of the film exhibits a change of $1^1/_2$ orders of magnitude. The observed response times for adsorption and desorption and oxygen gas sensitivity are as good as reported for sputtered ZnO films (Yen).

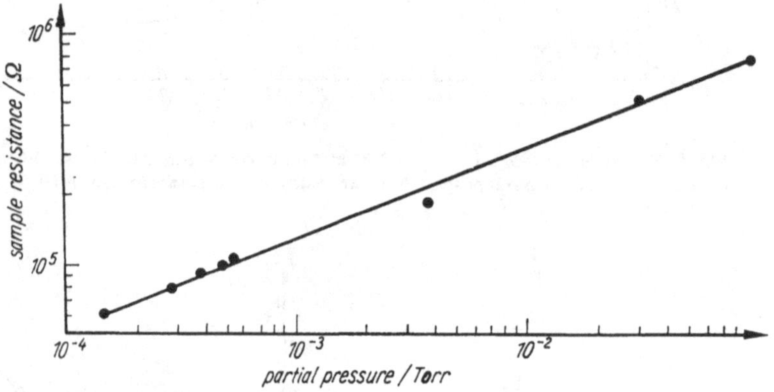

Fig. 6. Steady state resistance of the 15 h annealed ZnO thin film vs O_2 partial pressure at temperature 396 °C

Due to annealing of the film and increasing time of annealing the crystallinity improves (see Fig. 1). But the O_2 gas sensitivity for the films annealed upto 15 h is better than as-deposited and the films annealed upto 30 h. This means that the sensitivity has no unique dependence on the size of the grains. This type of the results have been reported also for sprayed SnO_2 thin films (Pink et al.). However, films annealed upto 30 h show large variation in resistance when they are held at 396 °C in vacuum. At this temperature, the sheet resistance are $\sim 1.1 \times 10^5$, 4.5×10^4, and 5.5×10^3 $\Omega \cdot cm^{-2}$ for the as-deposited, annealed 15 h and 30 h films, respectively. After introducing O_2 the resistance of the films annealed upto 30 h could not increase significantly (Fig.4). The times for adsorption and desorption for this film are large. It may be that for these films there is a sufficient oxygen loss from the lattice leading to a significant increase in the density of carriers (electrons).

In conclusion, we have prepared ZnO thin films by spray pyrolysis technique and these ZnO thin films exhibit promising properties in regard to reversible resistance variation with oxygen adsorption/desorption and hence these films can be used as low cost solid state gas sensors.

References

ADVANI, G. N., JORDAN, A. G.: J. Electronic Materials **9** (1980) 29
ARYA, S. P. S., D'AMICO, A., VERONA, E.: Thin Solid Films (to be published)
CHANG, S. C.: J. Vac. Sci. Tech. **A1**(2) (1982) 296
HEILAND, G.: Sensors and Actuators **2** (1982) 343
MORRISON, S. R.: Sensors and Actuators **2** (1982) 329
NITTA, M., HARADOME, M.: IEEE Trans. Electron Devices ED **26** (1979) 247
OYABI, T.: J. Appl. Physics **53** (1982) 7125
PINK, H., TRIETINGER, VITE, L.: Jap. J. Appl. Physics **19** (1979) 513
SETYAMA, T., KATO, K., FUJIISHI, K., NAGATANI, M.: Anal. Chem. **34** (1962) 1503
TANSLEY, T. L., NEELY, D. F.: Thin Solid Films **121** (1984) 95
WINDISCHMANN, H., MARK, P.: J. Electr. Chem. Soc. **126** (1979) 627
WEBB, J. B., WILLIAMS, D. F., BUCHANAM, M.: Appl. Phys. Lett. **79** (1981) 640
YEN, J. C.: J. Vac. Sci. Tech. **12** (1974) 47

(Received September 21, 1987;
accepted November 15, 1987)

Authors' address:

Dr. S. P. S. ARYA, Professor O. N. SRIVASTAVA
Thin Film and Electron Microscope Laboratory, Department of Physics
Banaras Hindu University
Varanasi — 221005, India

Book Review

I. Chikawa, K. Sumino, K. Wada (eds.)

Defects and Properties of Semiconductors: Defect Engineering

KTK Scientific Publishers Tokyo, D. Reidel Publishing Company, Dordrecht/Boston/Lancaster/Tokyo 1987, 261 S., ISBN 90-227-2352-4, Dfl. 230.00/ US $ 94.00.

Die Qualität der Halbleiterwerkstoffe bestimmt in modernen Basistechnologien zur Herstellung integrierter Schaltkreise die Ausbeute und Zuverlässigkeit entscheidend mit. Während in den zurückliegenden Jahren die Defektstruktur in Halbleiterwerkstoffen unter dem Aspekt der Defektunterdrückung untersucht wurde, steht neuerdings zunehmend die Kontrolle und Steuerung der Defektdichte und Defektverteilung im Vordergrund. Für dieses Forschungsgebiet setzt sich zunehmend der Begriff "Defect Engineering" durch.

Im Tagungsband, in dem 15 Vorträge des Symposiums "Defects and Qualities of Semiconductors" (Tokyo, 17.—18. Mai 1984) enthalten sind, wird versucht, diesem Trend sich einzuordnen. Im 1. Teil werden Verbindungshalbleiter (GaAs, InP, Heterostrukturen) behandelt, der 2. Teil ist dem einkristallinen Silicium gewidmet. Nachteilig ist die zeitliche Differenz zwischen dem Symposium und dem Erscheinen des Tagungsbandes. Gerade in den letzten Jahren sind große Fortschritte beim Übergang von der Beschreibung der Defektstruktur und den damit bedingten Materialeigenschaften zur Kontrolle und gezielten Beeinflussung erzielt worden. Ein Teil der in den Vorträgen enthaltenen Ergebnisse ist auch über andere Veröffentlichungen zugängig.

Dem Anliegen eines Defect Engineering wird im wesentlichen der 2. Teil gerecht. Behandelt werden das Verhalten von Punktdefekten in Siliciumkristallen, der Einfluß von Sauerstoff und Stickstoff auf die mechanischen und elektrischen Eigenschaften, die Wechselwirkung zwischen Versetzungen und Verunreinigungen sowie Grundlagen der Intrinsic-Getterung als eines der wirkungsvollsten Beispiele eines "Defect Engineering". Im Abschnitt zu den Verbindungshalbleitern stehen die Charakterisierung von Versetzungen, Fragen der Züchtung versetzungsfreier Kristalle, der Stöchometrie und die Charakterisierung tiefer Zentren im Vordergrund.

Der vorliegende Konferenzband vermittelt einen übersichtlichen Kenntnisstand über die Aktivitäten auf diesem Forschungsgebiet in Japan und ist eines der wenigen Ausgaben, die sich definitiv zu einem "Defect Engineering" bekennen. Die Beiträge der japanischen Autoren sind anspruchsvoll abgefaßt und zeichnen sich durch eine klare Darstellung aus. Umfassendes Datenmaterial wird angeboten. Der Band ist Festkörper- und Halbleiterphysikern, Werkstoffwissenschaftlern, Kristallographen und Halbleitertechnologen zu empfehlen.

H. Richter

Cryst. Res. Technol.	23	1988	5	677–681

V. RIEDE, N. SHARIF, H. NEUMANN, M. S. OMAR, H. SOBOTTA

Sektion Physik, Karl-Marx-Universität Leipzig, and
School of Physics, University of Bath

Infrared Optical Characterization of $(Cu_2GeSe_3)_{1-x}(CuGe_2P_3)_x$ Mixed Crystals

Infrared reflectivity spectra of $(Cu_2GeSe_3)_{1-x}(CuGe_2P_3)_x$ mixed crystals with $x = 0.1$ and 0.2 are measured in the wave-number range from 180 to 4000 cm^{-1}. An analysis of the spectra reveals contributions from five vibrational modes and from free carriers to the reflectivity. The results are compared with previous measurements for Cu_2GeSe_3 and $CuGe_2P_3$.

Infrarot-Reflexionsspektren von $(Cu_2GeSe_3)_{1-x}(CuGe_2P_3)_x$-Mischkristallen mit $x = 0.1$ und 0.2 wurden im Wellenzahlbereich von 180 bis 4000 cm^{-1} gemessen. Eine Analyse der Spektren ergibt Beiträge von fünf Schwingungsmoden und von freien Ladungsträgern zum Reflexionsvermögen. Die Ergebnisse werden mit früheren Messungen für Cu_2GeSe_3 und $CuGe_2P_3$ verglichen.

1. Introduction

Recently, the ternary semiconducting compound $CuGe_2P_3$ has attracted some attention and detailed studies of its electrical, lattice vibrational, thermal, and chemical bonding properties have been published (HAILING et al.; NEUMANN; NEUMANN et al., 1987a, b; PAMPLIN, OMAR; PRABHAKAR, SANYAL; RIEDE et al. 1984). Furthermore, it has been established that $CuGe_2P_3$ forms alloys with the elemental semiconductor germanium and the ternary compound Cu_2GeS_3, and some first experimental data regarding the physical properties of these mixed crystal systems have also been reported (BHIKSHAMAIAH et al. 1985, 1986; MACDONALD et al.; PAMPLIN, OMAR). In extending the mixed crystal studies with $CuGe_2P_3$ as one of the constituent compounds we found that alloys of the composition $(Cu_2GeSe_3)_{1-x}(CuGe_2P_3)_x$ can also be prepared. In the present work infrared optical measurements are used to obtain some first information on the lattice vibrational and electrical properties of these mixed crystals. The results are compared with related data for Cu_2GeSe_3 (RIEDE et al. 1987) and $CuGe_2P_3$ (PAMPLIN, OMAR; RIEDE et al. 1984).

2. Experimental

In complete analogy to the $(Cu_2GeS_3)_{1-x}(CuGe_2P_3)_x$ alloys (BHIKSHAMAIAH et al. 1986) the $(Cu_2GeSe_3)_{1-x}(CuGe_2P_3)_x$ mixed crystals were grown by the Bridgman method using a steep temperature gradient. Homogeneous samples suitable for the optical measurements could be prepared for the compositions $x = 0.1$ and 0.2. The samples used in the experiments were parallel-sided specimens cut from the Bridgman grown ingots. X-ray diffraction measurements showed that the material is polycrystalline and that the mixed crystals have the zincblende structure with a random distribution of the Cu and Ge atoms on the cation sites and the Se and P atoms on the anion sites. The same structural properties have been found in the $(Cu_2GeS_3)_{1-x}(CuGe_2P_3)_x$ mixed crystals, too (BHIKSHAMAIAH et al.

1985). The infrared reflectivity spectra of the samples were measured at room temperature in the wavenumber range $\bar{\nu} = 180-4000\ \text{cm}^{-1}$ using a Perkin-Elmer spectrophotometer model 580 B. None of the samples showed a measurable transmission. All the samples were p-type conducting as determined by hot point probe measurements.

3. Results and discussion

The measured reflectivity spectra for the wavenumber range $\bar{\nu} = 180-500\ \text{cm}^{-1}$ are shown in Figure 1, for $\bar{\nu} = 500-4000\ \text{cm}^{-1}$ in Figure 2. From Figure 1a it is obvious that the reflectivity spectrum of the sample with $x = 0.1$ exhibits a series of more or less pronounced structures which can be ascribed to lattice oscillators. The rapid increase of the reflectivity with decreasing frequency below about $300\ \text{cm}^{-1}$ is due to free carriers. In the sample with $x = 0.2$ the influence of free carriers is even more pronounced. Comparing Figures 1a and 1b we see that, firstly, the absolute values of the reflectivity are distinctly larger in the sample with $x = 0.2$ and, secondly,

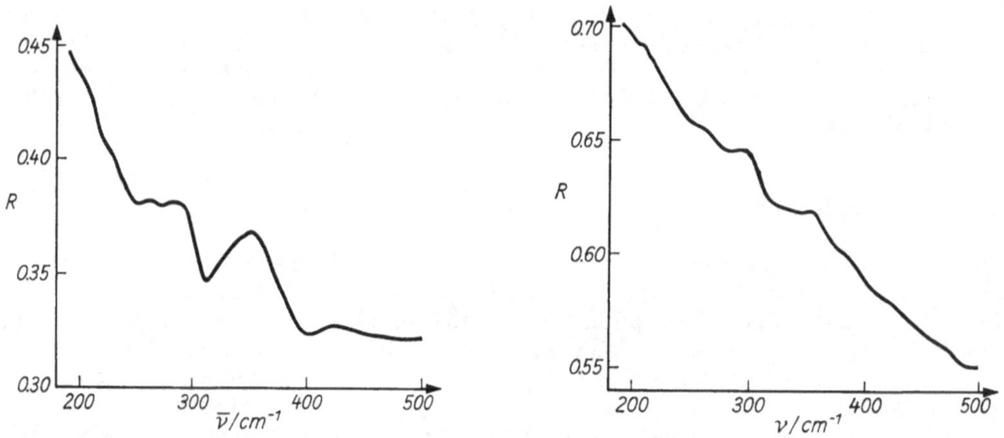

Fig. 1. Reflectivity spectra of $(Cu_2GeSe_3)_{1-x}(CuGe_2P_3)_x$ mixed crystals in the wavenumber range below $500\ \text{cm}^{-1}$ for $x = 0.1$ (a) and $x = 0.2$ (b)

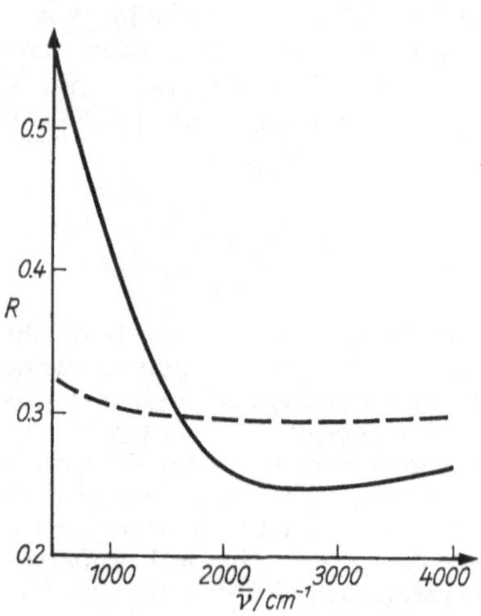

Fig. 2. Reflectivity spectra of $(Cu_2GeSe_3)_{1-x}(CuGe_2P_3)_x$ mixed crystals at wavenumbers $\bar{\nu} > 500\ \text{cm}^{-1}$ for $x = 0.1$ (dashed curve) and $x = 0.2$ (full curve)

as a consequence of the large free carrier contribution to the reflectivity spectrum the structures due to the lattice oscillators are less pronounced. Furthermore, it is evident from Figure 2 that the spectrum of the sample with $x = 0.2$ shows a broad but clearly pronounced reflectivity minimum at about 2700 cm^{-1} due to plasma reflection effects. The result found here for the sample with $x = 0.2$ is very similar to that obtained for $CuGe_2P_3$ (RIEDE et al. 1984) and indicates the presence of a rather high free carrier concentration.

Table 1
Frequencies of the transverse optical modes and free carrier parameters in $(Cu_2GeSe_3)_{1-x}(CuGe_2P_3)_x$ mixed crystals

Parameter	Cu_2GeSe_3	$x = 0.1$	$x = 0.2$
$\bar{\nu}_{TO}$ (cm^{-1})	202	205	203
	230	230	230
	267	265	265
	300	297	298
	—	362	363
pm_0/m_p (10^{20} cm^{-3})		1.6 ± 0.4	4.6 ± 0.3
$\mu m_p/m_0$ (cm^2/Vs)		2.1 ± 0.3	2.6 ± 0.1
ε_∞	12.6 ± 1.0	11.0 ± 1.0	9.5 ± 0.5
p (10^{20} cm^{-3})		1.0 ± 0.3	2.8 ± 0.2
μ (cm^2/Vs)		3.5 ± 0.4	4.3 ± 0.2

To determine the frequencies of the transverse optical vibrational modes the reflectivity spectra were first analysed with the Kramers-Kronig method. In this way we could identify five vibrational modes. The frequencies of these modes determined with an accuracy of about ± 3 cm^{-1} are compiled in Table 1 together with the vibrational mode frequencies found in Cu_2GeSe_3 (RIEDE et al. 1987). We see that the four low-frequency modes found in the mixed crystals correspond to those observed in Cu_2GeSe_3. Therefore, we conclude that these modes originate from vibrations in the Cu–Ge–Se subsystem of the mixed crystals. The mode with highest frequency occurs only in the mixed crystals and can therefore be ascribed to vibrations with the phosphorus atoms involved. On the basis of our data it can be concluded that the localized vibrational mode due to phosphorus atoms in Cu_2GeSe_3 has a frequency of about 360 cm^{-1}. We note that the transverse optical mode frequency of about 375 cm^{-1} found in $CuGe_2P_3$ (RIEDE et al. 1984) is only slightly higher than the frequency of the high-energy mode in the mixed crystals which confirms our supposition regarding the origin of this mode.

Using the vibrational mode frequencies obtained from the Kramers-Kronig analysis as starting parameters we have also determined the free carrier parameters of the samples. For this purpose, a complex dielectric function of the form

$$\varepsilon(\omega) = \varepsilon_\infty + \sum_{k=1}^{5} \frac{s_k \omega_k^2}{\omega_k^2 - \omega^2 - i\gamma_k\omega_k\omega} - \frac{\varepsilon_\infty \omega_p^2}{\omega^2 - i\omega/\tau} \tag{1}$$

was used to describe the reflectivity spectra. Here, ε_∞ is the high-frequency dielectric constant, s_k, ω_k and γ_k are the oscillator strength, the frequency, and the damping constant of the lattice modes, ω_p and τ are the plasma frequency and the relaxation time of the free carriers, and $\omega = 2\pi c\bar{\nu}$ is the angular frequency. Since the samples are p-type conducting the plasma frequency is given by

$$\omega_p^2 = \frac{e^2 p}{\varepsilon_v \varepsilon_\infty m_p} \tag{2}$$

where e is the elementary charge, p the hole concentration, ε_v the permittivity of free space and m_p the optical effective mass of the holes. Finally, if we assume that the relaxation time τ in relation (1) can be replaced by its average value $\langle\tau\rangle$ we can calculate the optical mobility μ of the holes according to the relation

$$\mu = e\langle\tau\rangle/m_p \,. \tag{3}$$

Now, in fitting relation (1) to the experimental reflectivity spectra we have determined ε_∞, ω_p and τ. Then, using relations (2) and (3) the values for the parameters pm_0/m_p and $\mu m_p/m_0$ are calculated and are also given in Table 1. To estimate the hole concentration p and mobility μ we need the effective mass m_p/m_0 of the holes (m_0 is the free electron mass). Unfortunately, no reliable Hall effect measurements could be made to determine the hole concentration independently because of the polycrystallinity of the samples. However, a reasonable estimate of the hole effective mass can be given.

In $CuGe_2P_3$ the hole effective mass has been determined experimentally from combined Hall effect and infrared reflectivity measurements (RIEDE et al. 1984) and has been found to be $m_p/m_0 = 0.56 \pm 0.10$. Within the given error limits this value agrees with the hole effective mass in GaP (BARANSKI et al.), the direct binary analog of $CuGe_2P_3$. Thus, it seems to be justified to suppose that the hole effective mass of Cu_2GeSe_3 is also approximately the same as that in ZnSe, the direct binary analog of this compound. For ZnSe a hole effective mass of $m_p/m_0 = 0.6$ has been reported (AVEN et al.) which differs only slightly from the effective hole mass in $CuGe_2P_3$. Therefore, we can certainly assume that the effective mass of the holes in the mixed crystals investigated is approximately equal to $m_p/m_0 = 0.6$ because of the small x values. On this assumption we obtain the hole concentrations and mobilities given in Table 1, too.

It is interesting to note that hole concentrations of the order of 10^{20} cm^{-3} have also been found in $CuGe_2P_3$ and in mixed crystals of $CuGe_2P_3$ with Ge (PAMPLIN, OMAR; RIEDE et al. 1984) but not in Cu_2GeSe_3 (RIEDE et al. 1987). Also the hole mobilities are in the same range as those in $CuGe_2P_3$ and in $(CuGe_2P_3)_{1-x}(6\,Ge)_x$ mixed crystals (PAMPLIN, OMAR). Therefore, it seems to be rather likely that the strongly pronounced p-type conductivity is due to the presence of $CuGe_2P_3$ in the mixed crystals. Furthermore, since all the samples are nominally undoped it must be concluded that the high hole concentrations are due to intrinsic defects. It cannot be excluded that similarly as in $CuGe_2P_3$ (PAMPLIN, OMAR) high concentrations of germanium atoms on anion sites are responsible for the p-type conductivity observed. If this interpretation is correct then it should be expected that other mixed crystal systems with $CuGe_2P_3$ as one of the constituent compounds are also p-type conducting and have high hole concentrations. Thus, to confirm this supposition it would be very interesting to extend the investigations to other mixed crystals of this type.

References

AVEN, M., MARPLE, D. T. F., SEGALL, B.: J. Appl. Physics **32** (1961) 2261

BARANSKI, P. I., KLOCHKOV, V. P., POTYKEVICH, I. V.: Semiconductor Electronics, Kiev 1975

BHIKSHAMAIAH, G., SARMA, M. V. S., SURYANARAYANA, S. V., OMAR, M. S.: J. Less-Common Metals **113** (1985) L5

BHIKSHAMAIAH, G., SURYANARAYANA, S. V., OMAR, M. S.: J. Mater. Sci. Lett. **5** (1986) 121

HAILING, T., SAUNDERS, G. A., OMAR, M. S., PAMPLIN, B. R.: J. Phys. Chem. Solids **45** (1984) 163

MacDonald, J. E., Saunders, G. A., Omar, M. S., Pamplin, B. R.: phys. stat. sol. (a) **89** (1985) K137
Neumann, H.: Cryst. Res. Technol. **20** (1985) K43
Neumann, H., Nowak, E., Omar, M. S.: Cryst. Res. Technol. **22** (1987a) 1437
Neumann, H., Nowak, E., Omar, M. S., Pamplin, B. R.: J. Less-Common Metals **132** (1987b) L13
Pamplin, B. R., Omar, M. S.: Progr. Crystal Growth Character. **10** (1985) 183
Prabhakar, N. V. K., Sanyal, S. P.: phys. stat. sol. (b) **137** (1986) 453
Riede, V., Neumann, H., Sobotta, H., Kissinger, W., Omar, M. S., Pamplin, B. R.: phys. stat. sol. (b) **123** (1984) K119
Riede, V., Sharif, N., Neumann, H., Sobotta, H., Omar, M. S., Pamplin, B. R.: Cryst. Res. Technol. **22** (1987) 1089

(Received and accepted November 3, 1987)

Author's addresses:

Dr. V. Riede, N. Sharif, Dr. H. Neumann, Dr. H. Sobotta
Sektion Physik, Karl-Marx-Universität
Linnéstraße 5
Leipzig
DDR-7010

Dr. M. S. Omar
School of Physics, University of Bath
Claverton Down
Bath BA2 7AY
England

Cryst. Res. Technol.	23	1988	5	682

Book Review

D. Walgraef (ed.)

Patterns, Defects and Microstructures in Nonequilibrium Systems – Applications in Materials Science

Proc. of an NATO Advanced Res. Workshop in Austin, Texas (USA), March 24–28, 1986.

Martinus Nijhoff Publ., Dordrecht, Boston, Lancaster 1987, with 377 p., numerous figs. Dfl. 195.00, US $ 85.50, UK £ 67.95, ISBN 90-247-3479-7.

Das einigende Band der teilnehmenden Physiker, Chemiker und Ingenieure waren die spontanen Ordnungsvorgänge in offenen Systemen fern vom thermodynamischen Gleichgewicht. Da inzwischen derartige Prozesse unmittelbare technische Bedeutung erlangt haben (z.B. Ionenimplantation, Laserstrahloberflächenvergütung und Elektronenstrahlheizung), verstärkt sich auch das Interesse an einem grundlegenden Verständnis des Geschehens im mikroskopischen Maßstab. Typische Nichtgleichgewichtskonzepte der irreversiblen Thermodynamik, der Bifurkationsanalyse und der nichtlinearen Reaktionsdiffusionsmechanik haben sich bereits in anderen Bereichen bewährt (z.B. Hydrodynamik, Biophysik) und verheißen Erfolg auch bei Anwendung auf den Festkörper. Denn dieser steht zweifellos im Mittelpunkt der 22 Originalbeiträge des vorliegenden Bandes, der von einem Aufsatz von I. Prigogine über ‚die Botschaft der Entropie' eingeleitet wird.

Die breite Palette der Themen sei durch folgende Stichworte gekennzeichnet: Dendritisches Wachstum, Strukturbildung während Konvektion, Quasikristalle, nichtkristalline Festkörper, zweidimensionale Phasen, chemische Wellen, Morphologie der Sedimentgesteine, Korrosion der Metalle, Laserschmelzen von Halbleitern, Stabilität bei Bestrahlung.

Den Schwerpunkt bilden aber 9 Arbeiten zur Mikroplastizität und zum Bruch mit besonderer Berücksichtigung der Instabilitäten des Versetzungsnetzwerks. Bereits bekannte und einige neue empirische Befunde werden anhand eines mehr oder weniger ausgefeilten Formalismus mit den Ausflüssen o. g. Konzepte konfrontiert und dadurch einer interessanten Prüfung unterzogen. So werden Zusammenhänge offenbart, die ohne Nichtgleichgewichtsanalyse wohl kaum erkannt worden wären. Naturgemäß bleibt noch vieles unerklärt. Aber Anregungen (evtl. zu weiterem eigenen Bemühen) kann sich der Leser allemal holen. Die Voraussetzungen an mathematische Fertigkeiten sind gering, eine gewisse thermodynamische Bildung ist wünschenswert. Literaturhinweise erleichtern den Zugang zu spezielleren Veröffentlichungen, die Angaben sind aber teilweise sehr lückenhaft. Das Buch ist vor allem jenen zu empfehlen, die sich mit dem Versetzungsverhalten bei Deformation und/oder ausgewählten Phasenbildungsprozessen beschäftigen.

P. Paufler

Cryst. Res. Technol. | 23 | 1988 | 5 | 683–690

V. G. ASLANIAN, H. P. BEZIRGANIAN, P. H. BEZIRGANIAN

Yerevan State University, Yerevan, USSR, and
Leninakan Branch of Yerevan Polytechnical Institute
Leninakan, USSR

Interference Pattern of X-Ray Waves inside the Crystal and its Image in Vacuum in Double-Field Approximation (I)

The distribution of intensity of the interference field during simultaneous superposition of all the four Bloch waves in a two-mode case has been investigated. It is shown that the distribution of intensity of the total field in the upper layers of the crystal is of a periodic character in two mutually-perpendicular directions. In the lower layers, where the first field is absorbed, a sinusoidal distribution occurs only in the direction of the reciprocal lattice vector.

Исследовано распределение интенсивности интерференционного поля при одновременной суперпозиции всех четырех блоховских волн в двухполевом случае. Показано, что распределение интенсивности суммарного поля в верхних слоях кристалла имеет периодический характер в двух взаимно перпендикулярных направлениях. В нижних слоях, где первое поле поглощается, синусоидальное распределение имеет место только в направлении вектора обратной решетки.

1. Introduction

As it is well-known (LAUE), during dynamic scattering of X-rays in a two-field (two-mode) case, four plane waves, two waves in each field, arise in a crystal (Fig. 1).

During investigations of the distribution of the intensity of Bloch waves in a crystal, the interference of two waves is usually considered: either between the waves propagated in the direction of point H of the reciprocal lattice (between $\boldsymbol{K}_{h1}$ and $\boldsymbol{K}_{h2}$), und between those propagated in the direction of point 0 of the reciprocal lattice

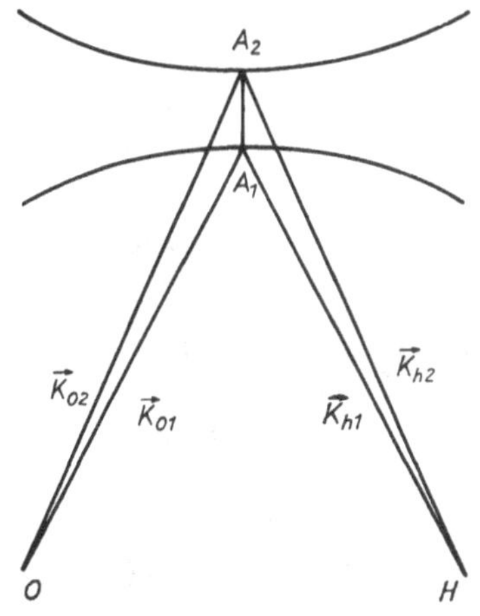

Fig. 1. Waves in a crystal in double-field approximation

(between $\boldsymbol{K}_{01}$ and $\boldsymbol{K}_{02}$), leading to a pendular distribution of the intensity in the crystal, or between the waves of the first field (between $\boldsymbol{K}_{01}$ and $\boldsymbol{K}_{h1}$) and between those of the second field ($\boldsymbol{K}_{02}$ and $\boldsymbol{K}_{h2}$), by which a physical interpretation is given to the weak absorption of the second field and to the strong absorption of the first field.

It has been shown in an early report (BEZIRGANIAN, BEZIRGANIAN 1983) that in a crystal a separate examination of the interference of two out of the four coherent waves is generally incorrect. In thin crystals (non-absorbing), in particular, the true distribution of the intensity of the common field has a rather complex form (Fig. 2).

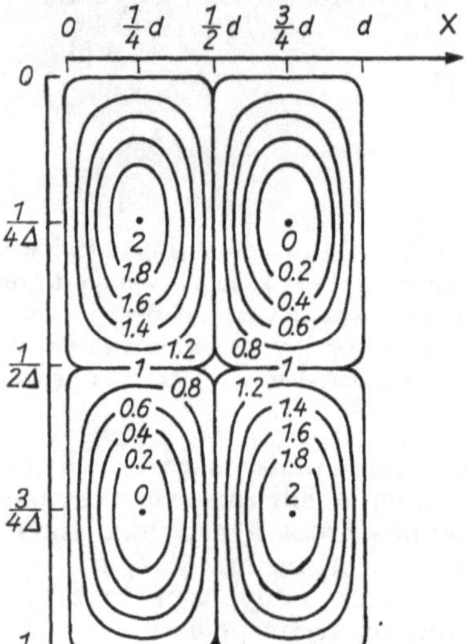

Fig. 2. Distribution of intensity in a non-absorbing crystal, in case when the primary beam falls exactly at a Bragg angle. d is the interplanar space (Laue symmetrical case)

In the present study we have investigated also the distribution of intensity between adjacent reflecting planes in the direction of the diffraction vector. A number of reports (KATO 1968a, b, c) deal with the distribution of the intensity of diffracted waves in spherical wave approximation inside and outside the crystals, depending on the thickness and the absorption of the diffracting crystal.

The above-mentioned reports, however, separately examine the interference of waves propagated in the direction of point 0 and of those propagated in the direction of point H of the reciprocal lattice; moreover, the intra-crystal distributions of intensity are not always compared with the extra-crystal ones and the conditions of observation of intra-crystal propagations are not considered in detail. These features have been dealt with in the present investigation.

2. Investigation of the distribution of intensity in the interference field of Bloch waves inside the crystal

In the most general case, the distribution of intensity in the interference field of the four Bloch waves

$$\mathcal{D}_{01} = \tfrac{1}{2}\, \mathcal{D}_0^i \exp\,(i\, 2\pi \boldsymbol{K}_{01}\boldsymbol{r}) \tag{1}$$

$$\mathcal{D}_{02} = \tfrac{1}{2}\, \mathcal{D}_0^i \exp\,(i\, 2\pi \boldsymbol{K}_{02}\boldsymbol{r}) \tag{2}$$

$$\mathcal{D}_{h1} = \tfrac{1}{2}\, \mathcal{D}_0^i \exp\,(i\, 2\pi \boldsymbol{K}_{h1}\boldsymbol{r}) \tag{3}$$

$$\mathcal{D}_{h2} = -\tfrac{1}{2}\, \mathcal{D}_0^i \exp\,(i\, 2\pi \boldsymbol{K}_{2h}\boldsymbol{r}) \tag{4}$$

for a double-field plane-wave approximation is expressed by the formula

$$I = |\mathcal{D}_{01} + \mathcal{D}_{02} + \mathcal{D}_{h1} + \mathcal{D}_{h2}|^2$$

which, for a symmetric case of reflection and at σ polarization, taking into account (1)–(4) and the following correlations

$$\boldsymbol{K}_{h1} - \boldsymbol{K}_{01} = \boldsymbol{H}, \quad \boldsymbol{K}_{h2} - \boldsymbol{K}_{h1} = \boldsymbol{A}_2\boldsymbol{A}_1, \quad \boldsymbol{K}_{02} - \boldsymbol{K}_{01} = \boldsymbol{A}_2\boldsymbol{A}_1$$

$$\boldsymbol{K}_{h2} - \boldsymbol{K}_{02} = \boldsymbol{H}, \quad \boldsymbol{K}_{h1} - \boldsymbol{K}_{02} = \boldsymbol{H} - \boldsymbol{A}_2\boldsymbol{A}_1, \quad \boldsymbol{K}_{h2} - \boldsymbol{K}_{01} = \boldsymbol{H} + \boldsymbol{A}_2\boldsymbol{A}_1$$

may be brought to

$$I = |\mathcal{D}_0^i|^2 \left[+ \sin(2\pi \boldsymbol{H}\boldsymbol{r}) \sin(2\pi \boldsymbol{A}_2\boldsymbol{A}_1\boldsymbol{r})\right] \tag{5}$$

where $|\mathcal{D}_0^i|^2$ is the intensity of the primary wave.

The latter, allowing for the absorption (Bezirganian, Bezirganian 1985), will come to the following form:

$$I(x, z) = |\mathcal{D}_0^i|^2 \left\{\exp\left[-(\mu + \sigma)\frac{z}{\gamma_0}\right] \cos^2\left(\frac{\pi x}{d}\right) + \right.$$
$$+ \exp\left[-(\mu - \sigma)\frac{z}{\gamma_0}\right] \sin^2\left(\frac{\pi x}{d}\right) +$$
$$\left. + \exp\left(-\mu\frac{z}{\gamma_0}\right) \sin\left(\frac{2\pi x}{d}\right) \sin(2\pi z\Delta)\right\} \tag{6}$$

where

$$\Delta = \frac{k|\chi_{hr}|}{\gamma_0} \qquad \gamma_0 = \cos\theta_B$$
$$\mu = 2\pi K\,|\chi_{0i}| \qquad \sigma = 2\pi K\,|\chi_{hi}|$$
$$\chi_0 = -|\chi_{0r}| - i\,|\chi_{0i}|$$
$$\chi_h = |\chi_{hr}| \exp(i\zeta_h) + i\,|\chi_{hi}| \exp(i\omega_h)$$
$$\chi_{\bar{h}} = |\chi_{hr}| \exp(-i\zeta_h) + i\,|\chi_{hi}| \exp(-i\omega_h)$$
$$0 \le \zeta_h \le \pi;\ 0 \le \omega \le \pi$$

χ_0, χ_h, $\chi_{\bar{h}}$ are Fourier's components of polarizability, K is the wave number of the incident wave, the axis z is directed normal to the surface of the crystal (inward), whereas the axis x is directed to the vector $\boldsymbol{H}$ of the reciprocal lattice.

The distribution of intensity inside the non-absorbing crystal according to (5) is shown in Figure 2. Figures 3–5 represent the distribution of intensity in the absorbing crystal at various depths between the adjacent planes. The curves have been drawn up for a silicon crystal, reflection (220), radiation CuK_{α_1}. It is obvious from these figures that, as we go deeper in the crystal, the distribution of intensity of Bloch waves acquires the form of a distribution of intensity in the interference field of the second field; the first field ($\boldsymbol{K}_{01}$ and $\boldsymbol{K}_{h1}$) is absorbed and only an interference between the $\boldsymbol{K}_{02}$ and $\boldsymbol{K}_{h2}$ waves of the second field is observed.

As it is evident from (6), the intensity of the common field in the absorbing media consists of three members, which are diminished as the expressions

$$\exp\left\{-(\mu + \sigma)\frac{z}{\gamma_0}\right\}, \qquad \exp\left\{-(\mu - \sigma)\frac{z}{\gamma_0}\right\}, \qquad \exp\left\{-\mu\frac{z}{\gamma_0}\right\},$$

respectively. As we go deeper in the crystal, the first to disappear is the first (strongly absorbable) member, the cosinusoidal distribution and then the third member having the usual absorption coefficient. Finally, in the depths of thick absorbing crystals

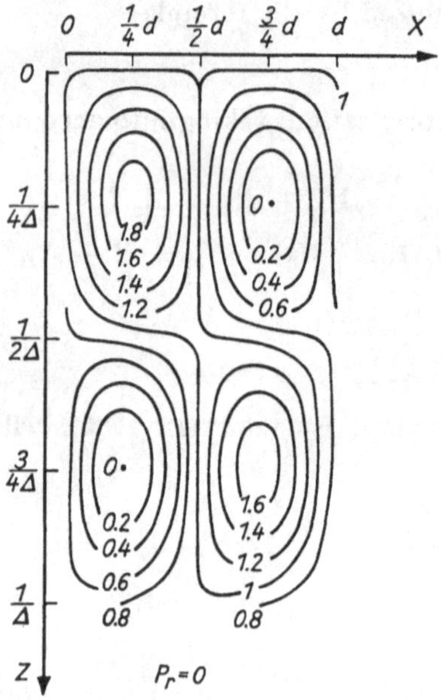

Fig. 3. Distribution of intensity in an absorbing crystal (when $0 \leq Z \leq 1/\Delta$)

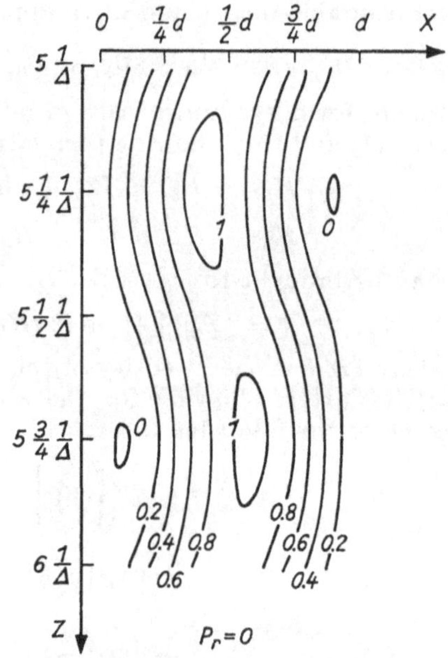

Fig. 4. Distribution of intensity in case $5 \cdot 1/\Delta \leq Z \leq 6 \cdot 1/\Delta$

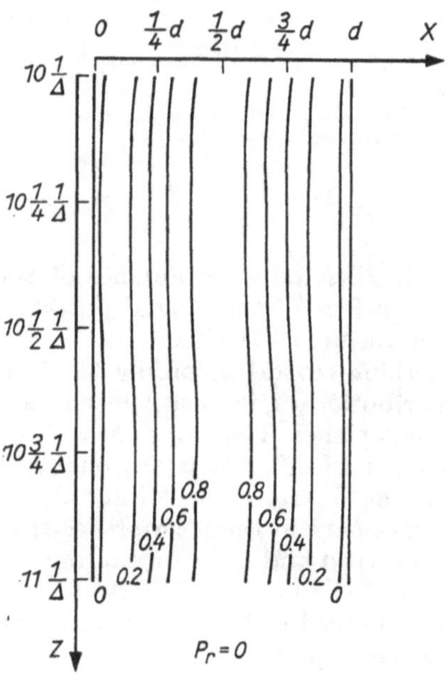

Fig. 5. Distribution of intensity in case $10 \cdot 1/\Delta \leq Z \leq 11 \cdot 1/\Delta$

remains the second (weakly absorbable) member, the sinusoidal distribution of intensity.

Thus, we come to the following conclusions:

1. The intensity of the interference field of all the four Bloch waves in a two-mode case in the crystal does not contain a purely pendular distribution.

2. In thick absorbing samples, as we go deeper in the crystal, a sinusoidal distribution in the direction of the reciprocal lattice vector sets in only after the disappearance of the first and third members (after the absorption of the first field) (Fig. 5).

3. Observation of the distribution of intensity of Bloch waves outside the crystal

Of great interest is the investigation of the mechanism of the emergence of fringe patterns of Bloch waves outside the crystal and the relation between the intra- and extra-crystal patterns.

3.1. Mechanism of the emergence of pendular bands outside the crystal

As we were convinced by the use of (5) and (6), the fringe pattern of the four Bloch waves in a two-mode case in the crystal does not contain a purely pendular distribution of intensity.

Then arises the question: How are pendular bands observed outside a wedge-shaped crystal?

Before answering this question, let us discuss the conception existing about this matter. We have already pointed out that a separate examination of the interference between the $\boldsymbol{K}_{01}$ and $\boldsymbol{K}_{02}$ waves and between $\boldsymbol{K}_{h1}$ and $\boldsymbol{K}_{h2}$ waves leads to a pure pendular distribution of intensity in the crystal (Fig. 1). In this approximation, for the intensities of I_0 and I_h waves diffracted in the direction of incidence and for the waves diffracted in the direction of reflection, we obtain

$$I_0 = |\mathcal{D}_{01} + \mathcal{D}_{02}|^2 = |\mathcal{D}_0^i|^2 \cos^2 (\pi \boldsymbol{A}_2 \boldsymbol{A}_1 \boldsymbol{r}) , \tag{7}$$

$$I_h = |\mathcal{D}_{h1} + \mathcal{D}_{h2}|^2 = |\mathcal{D}_0^i|^2 \sin^2 (\pi \boldsymbol{A}_2 \boldsymbol{A}_1 \boldsymbol{r}) . \tag{8}$$

The expressions (7) and (8) lead to the following conclusions:

1. Pendular bands are not obtained from plane-parallel plates. Depending on the thickness of the crystal, the diffracted wave must emerge from it in the direction of incidence or in the direction of reflection, a fact which is not observed experimentally, supposedly because of the roughness of the exit surface.
2. The use of a wedge-shaped crystal is necessary for obtaining pendular bands, a fact, which again, as we shall see below, is not obligatory.

In our subsequent investigations we have to proceed from the following general principle: Only those distributions of intensity arise outside the crystal, for which precisely there conditions for their appearance are realized, irrespective of intra-crystal distributions, that is, a given distribution may be absent in the crystal, but may arise outside it and vice versa.

In fact, for the appearance of pendular bands it is necessary to superimpose at a small angle (of the order of seconds) in vacuum (outside the crystal) waves having wave vectors $\boldsymbol{K}_{01}^t$ and $\boldsymbol{K}_{02}^t$ or waves having wave vectors $\boldsymbol{K}_{h1}^t$ and $\boldsymbol{K}_{h2}^t$. In case of a plane-parallel crystalline plate in vaccum these waves are strictly parallel to each other in pairs and therefore pendular bands are not obtained from plane-parallel plates. The superposition of the waves of the second field ($\boldsymbol{K}_{02}$ and $\boldsymbol{K}_{h2}$) is realized in thick samples (Fig. 7), but outside the crystal this phenomenon occurs only at negligible distances from the crystal. Thus, a periodic distribution of intensity in the direction of the reciprocal lattice in thick crystals occurs inside the crystal and is virtually absent outside it.

3.2. Observation of pendular bands from a parallelepiped crystal

We have already mentioned that for the observation of pendular bands outside the crystal it is necessary to create conditions for the superposition of waves having wave vectors $\boldsymbol{K}_{01}^t$ and $\boldsymbol{K}_{02}^t$, or of those having wave vectors $\boldsymbol{K}_{h1}^t$ and $\boldsymbol{K}_{h2}^t$, provided that the angles between these wave vectors and the amplitude differ from zero but not by much.

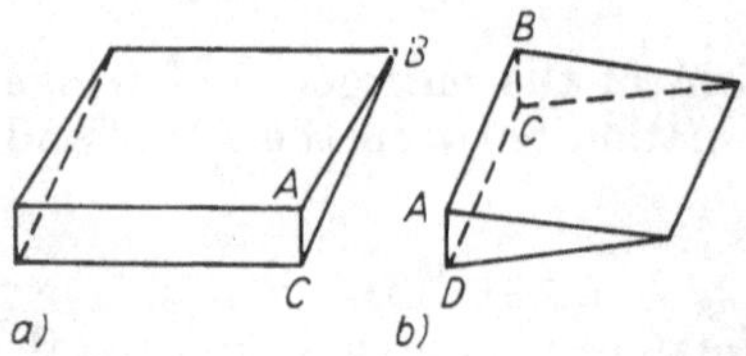

Fig. 6. Reflecting planes of wedge-shaped crystals ABCA (a) and ABCDA (b)

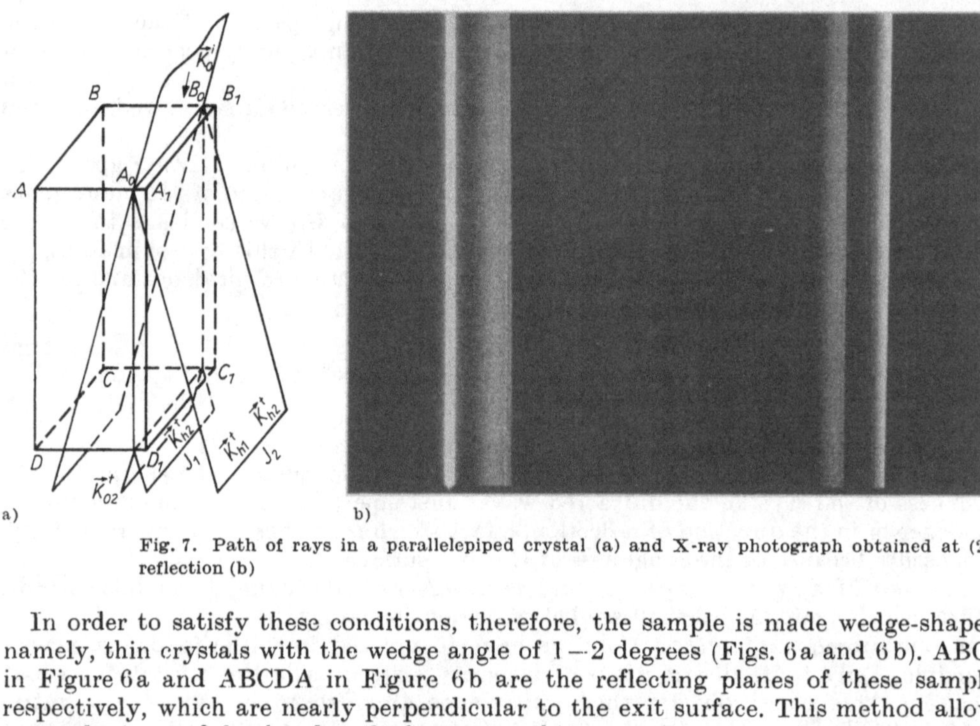

a) b)

Fig. 7. Path of rays in a parallelepiped crystal (a) and X-ray photograph obtained at (220 reflection (b)

In order to satisfy these conditions, therefore, the sample is made wedge-shaped, namely, thin crystals with the wedge angle of 1–2 degrees (Figs. 6a and 6b). ABCA in Figure 6a and ABCDA in Figure 6b are the reflecting planes of these samples respectively, which are nearly perpendicular to the exit surface. This method allows us to obtain pendular bands only from very thin crystals.

It appears that, by using the method of reduction of the path of diffracted waves in the crystal, it is possible to obtain pendular bands from thick samples, where the angle between the exit surfaces and reflecting planes is equal to that of Bragg or less.

Here we shall discuss a particular case for obtaining pendular bands from a parallelepiped crystal, where the angle between the reflecting plane and the exit surface is equal to zero, i.e., waves having wave vectors $\boldsymbol{K}_{h1}^t$ and $\boldsymbol{K}_{h2}^t$ emerge from the lateral surface (Fig. 7a). The planes ABCDA || $A_1B_1C_1D_1A_1$ are reflecting planes. A parallelepiped sample (Fig. 7a) has been prepared from a dislocation-free sample. A sheet beam falls on the reflecting planes of the crystal and pendular bands are obtained in the form of parallel lines. If the primary wave falls on the crystal not far from the edge A_1B_1 in order to avoid the absorption of the first field, then pendular bands, as shown in Figure 7b, are obtained.

We may be convinced in this fact by a graphical investigation of the disposition of the wave vectors inside and outside the crystal (Fig. 8).

The primary flat wave $\boldsymbol{K}_0^i$ falls on the reflecting planes (hkl) at an exact Bragg angle. Inside the crystal, the propagation points A_1 and A_2 and the waves having wave vectors $\boldsymbol{K}_{h1}$, $\boldsymbol{K}_{h2}$, $\boldsymbol{K}_{01}$ and $\boldsymbol{K}_{02}$ arise by the intersection of the normal LN_0 of the entrance surface with the dispersive surfaces.

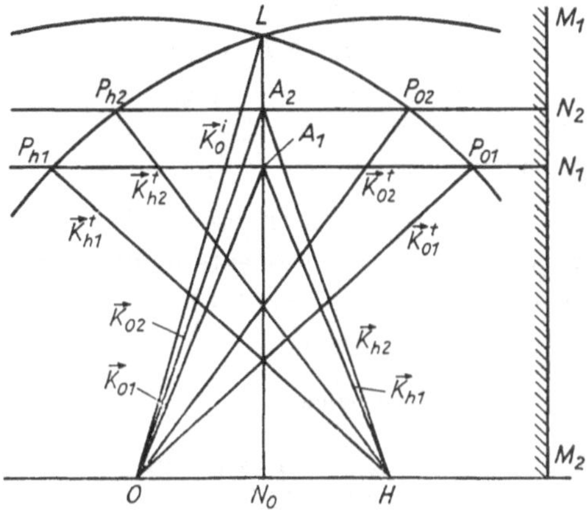

Fig. 8. Disposition of wave vectors in the crystal shown in Figure 7a

Since the waves emerge from the lateral surfaces (the first two from the right and the last two from the left) then the common normals A_1N_1 and A_2N_2 to the lateral surface M_1M_2 pass through the propagation points A_1 and A_2. As a result of the intersection of these normals with the spheres of incidence and reflection, the points of incidence P_{01} and P_{02} and the points of reflection P_{h1} and P_{h2} arise. By joining these points correspondingly with the points of the reciprocal lattice O and H, the wave vectors outside the crystal $\boldsymbol{K}_{01}^t$, $\boldsymbol{K}_{02}^t$, $\boldsymbol{K}_{h1}^t$ and $\boldsymbol{K}_{h2}^t$ are obtained. It may be seen in Figure 8 that pendular bands with a period

$$\Delta_h = \frac{1}{|\boldsymbol{K}_{h1}^t - \boldsymbol{K}_{h2}^t|} = \frac{1}{|\boldsymbol{P}_{h1}\boldsymbol{P}_{h2}|} = \frac{\sin\theta}{|\boldsymbol{A}_2\boldsymbol{A}_1|}$$

arise as a result of the superposition of the waves having wave vectors $\boldsymbol{K}_{h1}^t$ and $\boldsymbol{K}_{h2}^t$.

As it is generally known, the waves of the first field having wave vectors $\boldsymbol{K}_{01}$ and $\boldsymbol{K}_{h1}$ are strongly absorbed, considering that the wave with a wave vector $\boldsymbol{K}_{01}$ passes a long way through the crystal (Fig. 8) and is almost completely absorbed, consequently, pendular bands are not observed outside the crystal in the direction of the point of the reciprocal lattice and that the wave with a wave vector $\boldsymbol{K}_{h1}$ passes a short way through the crystal, i.e., it is slightly depressed and therefore pendular bands are observed in the direction of the point H of the reciprocal lattice (Fig. 7b).

4. Conclusions

In the approximation of the simultaneous interference of all the four Bloch waves in a two-mode case, it has been demonstrated that:

1. In the upper layers of the crystal adjoining the entrance surface in which the strongly absorbable (first) field is insignificantly weakened and the interference occurs

by the superposition of all the four waves, the distribution of intensity in the interference field is expressed by the formula (5). As it is obvious from this expression, the intensity of the common field has a periodic distribution simultaneously in two mutually-perpendicular directions, in the direction of the normal to the entrance surface and in the direction of the reciprocal lattice vector of the reflecting planes; meanwhile, as it is evident from (7) and (8) during a separate examination of the interference between the waves having wave vectors $\boldsymbol{K}_{01}$ and $\boldsymbol{K}_{02}$ and between those having wave vectors $\boldsymbol{K}_{h1}$ and $\boldsymbol{K}_{h2}$, we obtain a uniform distribution for the intensity of the common field

$$I = I_0 + I_h = |\mathscr{D}_0^i|^2 .$$

2. In the deeper layers of the crystal, where the first field is completely absorbed and the weakening of the second field may be disregarded, a sinusoidal distribution in the direction of the reciprocal lattice vector is obtained for the total intensity

$$I = |\mathscr{D}_{01} + \mathscr{D}_{02} + \mathscr{D}_{h1} + \mathscr{D}_{h2}|^2 = |\mathscr{D}_{02} + \mathscr{D}_{h2}|^2 = |\mathscr{D}_0^i|^2 \sin^2 (\pi \boldsymbol{H}\boldsymbol{r})$$

as in the case of a separate examination of the supreposition of waves having wave vectors $\boldsymbol{K}_{02}$ and $\boldsymbol{K}_{h2}$ (Fig. 5).

References

Bezirganian, H. P., Bezirganian, S. E.: Uchonie Zapiski Yerevan State University, Physika **2** (1983) 153

Bezirganian, H. P., Bezirganian, S. E.: Cryst. Res. Technol. **20** (1985) 1

Kato, N.: J. Appl. Physics **39** (1968) 2281 a

Kato, N.: Acta geologica Bratislava **14** (1968) 43 b

Kato, N.: J. Appl. Physics **39** (1968) 225 c

Laue, M.: Röntgenstrahlinterferenzen, Frankfurt/Main 1960

(Received October 23, 1987; accepted October 30, 1987)

Authors' addresses:

Prof. Dr. sc. P. H. Bezirganian, H. P. Bezirganian
Yerevan State University, Chair of Solid State Physics
Mravyan 1
375049 Yerevan 49
Armenia USSR

Dr. V. G. Aslanian
Leninakan Branch of Yerevan Polytechnical Institute
Savoyan 7
377500 Leninakan
Armenia USSR

Cryst. Res. Technol.	23	1988	5	691—697

A. M. RADWAN

Theoretical Physics Lab., National Research Centre
Cairo, Egypt

Structural Energetics of the Intermetallic Compounds $CaMg_2$ and $CaAl_2$

The structures of the Laves phases $CaMg_2$ and $CaAl_2$ are investigated using the optimized model potential. Charge transfer, which is taken into account in an approximate and simple way, plays an important role in the determination of the stable crystal structures of both $CaMg_2$ and $CaAl_2$.

Die Strukturen der Lavesphasen $CaMg_2$ und $CaAl_2$ werden mit Hilfe des optimierten Modellpotentials untersucht. Umladungseffekte, die in einer einfachen Weise annähernd erfaßt werden, spielen eine wichtige Rolle in der Bestimmung der stabilen Kristallstrukturen von $CaMg_2$ und $CaAl_2$.

1. Introduction

The application of pseudo or model potentials in the theory of simple metals was connected with an essential progress in understanding the properties of these metals (HEINE, WEAIRE). Especially it is possible to calculate the observed stable structures by comparing the energy of different structures and looking for the lowest one. These calculations are successful for pure metals (RADWAN; RADWAN, SALEH).

The question arises, whether it is possible to extend these investigations to intermetallic compounds (IMC), but only few results are available (INGLESFIELD; MROSAN et al.; RENNERT, RADWAN). It should be borne in mind that there are some additional complications for IMC compared with pure metals. Small reciprocal lattice vectors lower the convergence of the perturbation series. There are stronger deviations of the charge density from a constant value which may not be included in linear screening. This means that charge transfer plays an important role. Furthermore, the situation is complicated by the large differences in the electrostatic energies E_{es} of the structures of the compounds to be compared (RADWAN(1)), contrary to the case of simple metals where E_{es} of fcc, bcc and hcp structures is nearly the same. Thus it is of interest to investigate to what extent the pure metal scheme for the calculation of the structural energy E_{st} in second order perturbation theory is applicable to intermetallic compounds. We have already investigated this question in the case of the Laves phase $MgZn_2$ (RENNERT, RADWAN) where both Mg and Zn have the same valence 2. But in the present paper we have a further difficulty arising from the fact that in $CaAl_2$, Ca and Al have different chemical valencies, 2 and 3, respectively. This means a further departure from linear screening than in case of $MgZn_2$. The Laves phases are the largest group of IMC and there is some evidence that the structures C14 and C15 (SCHULZE) are electronic phases like hcp and fcc for pure metals. C14 and C15 are close-packed structures of two sorts of spheres with ideal diameter ratio $\sqrt{3/2}$ (SCHULZE), analogous to hcp and fcc structures for one sort of spheres. The fact that model potential theory is applicable to simple metals only but not to transition metals makes a drastic limitation on its use for IMC. There are only few Laves phases consisting of two simple metals.

2. Structural energy of compounds and binary alloys

The structural energy per ion E_{st} is the sum of two terms (HEINE, WEAIRE) (at temperature $T = 0$ and a fixed volume)

$$E_{st} = E_{es} + E_{bs} \tag{1}$$

E_{es} is the electrostatic energy and E_{bs} is the band structure energy, which has to be calculated to second order perturbation using Shaw's optimized model potential (SHAW 1968). The Laves phases C14 ($MgZn_2$) and C15 ($MgCu_2$) may be considered as a stacking sequence of three different types of layers as described by RADWAN (1): Ca planes (α, β, γ), MgI (or AlI) planes (a, b, c) and MgII (or AlII) Kagomé nets (A, B, C). A certain structure is characterized by noting the positions of the Kagomé nets only. For example C14 $= \beta B\beta a\gamma C\gamma a$ = BC, C15 $= \alpha A\alpha c\beta B\beta a\gamma C\gamma b$ = ABC. Therefore, in the structural energy of a structure J,

$$E_{st}(J) = \sum_g{}' E_g(J) + \text{Const.} \qquad E_g(J) = \sum_{M,M'} S_M^J(\boldsymbol{g})\, S_{M'}^{*J}(g)\, f_{MM'}(\boldsymbol{g}) \tag{2}$$

The summations are carried out over planes at first and over a special stacking sequence secondly as described by RADWAN (1). $S_M^J(g)$ is the structure factor of atoms of sort M (M = Ca, MgI or AlI, MgII or AlII) and $f_{MM'}(g)$ is given by

$$f_{MM'}(\boldsymbol{g}) = \frac{2\pi}{Vg^2} Z_M Z_{M'}\,, \left(e^{-g^2/4\lambda} - \overline{F}_{MM'}^N(\boldsymbol{g})\right). \tag{3}$$

V is the volume per ion, Z_M and $Z_{M'}$ are the effective valencies (or charges) of atoms of type M and M', respectively, λ is the parameter of the Gaussian caps and

$$\overline{F}_{MM'}^N(q) = \tfrac{1}{2}\left[F_{MM'}^N(q) + F_{M'M}^N(q)\right] \tag{4}$$

$F_{MM'}^N(q)$ is the normalized energy-wave number characteristic which is defined analogous to that of simple metals (SHAW 1969) as (in Hartree approximation)

$$F_{MM'}^N(q) = \frac{-q^2V^2}{4\pi Z_M Z_{M'}} \int\limits_{k \le k_F} d^3k\, \frac{w_M(\boldsymbol{k}, \boldsymbol{q})\, u_{M'}^0(\boldsymbol{k}, \boldsymbol{q})}{k^2 - |\boldsymbol{k} + \boldsymbol{q}|^2}\,. \tag{5}$$

Expressions for the screened form factor $w_M(\boldsymbol{k}, \boldsymbol{q})$ of atoms of type M and the unscreened one $u_{M'}^0(\boldsymbol{k}, \boldsymbol{q})$ of the M'-atoms are identical with those given by SHAW (1968). Corrections due to exchange and correlation are then added, however, our treatment is somewhat different from that of SHAW (1970). The final formula for $F_{MM'}^N$ is given by RADWAN and RENNERT. Equation (2) has then the form:

$$E_g(J) = \frac{1}{N^2} Re \sum_{g\perp} \sum_{M=1}^{3} \sum_{M'=M}^{3} \sum_{L=1}^{3} \sum_{L'=k}^{3} (2 - \delta_{MM'}\delta_{LL'})\, SM(J, L, \boldsymbol{g}_\perp)$$

$$SM^*(J, L', \boldsymbol{g}_\perp)\, W_{MM'LL'}(\boldsymbol{g}_\perp) \tag{6}$$

where in the summation over L': k = L if M = M' and k = 1 otherwise and

$$W_{MM'LL'}(\boldsymbol{g}_\perp) = \sum_{g_{||}} S1(L, \boldsymbol{g}_{||})\, S^*1(L', \boldsymbol{g}_{||})\, S2(M, L, \boldsymbol{g}_{||})\, S^*2(M', L', \boldsymbol{g}_{||})$$

$$f_{MM'}\sqrt{(\boldsymbol{g}_{||}^2 + \boldsymbol{g}_\perp^2)} \tag{7}$$

$S1(\mathrm{L}, \boldsymbol{g}_{||})$ is the structure factor of the inversion centre of the layer with position L. $S2(\mathrm{M}, \mathrm{L}, \boldsymbol{g}_{||})$ is the structure factor of the layer in the position L and containing atoms of type M.

$SM(\mathrm{J}, \mathrm{L}, \boldsymbol{g}_{\perp})$ is the structure factor of the sequence of layers with position L and atom type M (RADWAN (1)). It should be noticed that both structure factors $S1(\mathrm{L}, \boldsymbol{g}_{||})$ and $S2(\mathrm{M}, \mathrm{L}, \boldsymbol{g}_{||})$ are independent on the structure under consideration, but $SM(\mathrm{J}, \mathrm{L}, \boldsymbol{g}_{\perp})$ is a structure dependent one (RADWAN (1)).

3. Results and discussion

The experimentally observed structure of $CaMg_2$ is C14 = BC, whereas that of $CaAl_2$ is C15 = ABC. The alloy values of k_F are: $k_F(CaMg_2) = 0.67763$ and $k_F(CaAl_2) = 0.81649$, corresponding to atomic volumes $V = 190.32$ and 145.06 a.u., respectively. We have calculated $\overline{F}^N_{MM'}$ for $CaMg_2$ and $CaAl_2$ from 0 to $6k_F$ at intervals TQ $= 0.04\,k_F$ using the exchange and correlation function of (SINGWI et al.) and the model potential parameters of APPAPILLAI and HEINE for Ca, Mg, and Al. However, the use of model potential parameters of pure Ca, Mg, and Al in the calculations concerning $CaMg_2$ and $CaAl_2$ is justified only by assuming that the change in the model potential parameters of pure metals upon alloying is very small, like $MgZn_2$ (less than 2%) (RENNERT et al.). Obviously, the effective charges Z_{Ca}, Z_{Mg}, and Z_{Al} must be recalculated at $k_F(CaMg_2)$ and $k_F(CaAl_2)$, because the change in the depletion charge on alloying is large (TAUT, PAASCH). The calculated alloy values are:

$Z_{Ca} = 2.36250$ and $Z_{Mg} = 2.13948$ for $CaMg_2$, and

$Z_{Ca} = 2.52695$ and $Z_{Al} = 3.17708$ for $CaAl_2$.

We have calculated the structural energy E_{st} for all possible but non-equivalent stacking variants consisting of 2, 3, and 6 layers, namely, the four structures:

2 layers: CB = C14
3 layers: ABC = C15
6 layers: BABCAC = 6a
6 layers: ABACAC = 6b

The calculations are carried out using the ideal values of:

1) the axial ratio: $c/a = \sqrt{8/3}$
2) the positions x of MgII (or AlII) in Kagomé planes: $x = 1/6$ and
3) the parameter z giving the heights of the Ca atoms: $z = 1/16$ (PEARSON).

The electrostatic energy E_{es} is calculated with $\lambda = 0.5$ and given in Table 1 in a.u. From Table 1 it is clear that E_{es} favours C15 structure over 6a, 6b, and C14 structures. The difference $\Delta E_{es} = |E_{es}(C15) - E_{es}(C14)|$ is quite large, so that for instance to obtain C14 as the most stable structure for $CaMg_2$, this ΔE_{es} must be outbalanced

Table 1
E_{es} (a.u.) of the different structures of $CaMg_2$ and $CaAl_2$

Structure	E_{es} ($CaMg_2$)	E_{es} ($CaAl_2$)
C15	−1.223115	−2.373012
6a	−1.222973	−2.372447
6b	−1.222830	−2.371882
C14	−1.222686	−2.371314

by the difference $\Delta E_{bs} = |E_{bs}(C14) - E_{bs}(C15)|$, i.e. $\Delta E_{bs} > \Delta E_{es}$ which is rather difficult to achieve without taking charge transfer into consideration. The structural energy E_{st} calculated without charge transfer for the different structures under consideration are given in Table 2. A cubic interpolation scheme is used to compute E_{bs} from the tabulated values of the energy wave number characteristic $\bar{F}^{N}_{MM'}(q)$. Unlike pure metals (RADWAN, SALEH), it is unpractical to calculate $\bar{F}^{N}_{MM'}(q)$ directly at the reciprocal lattice vectors $\boldsymbol{g}$, because the number of $\boldsymbol{g}$'s in IMC is far greater than in case of pure metals. That is why interpolation has to be used here.

Table 2
E_{st} (a.u.) of the different structures of $CaMg_2$ and $CaAl_2$

Structure	E_{st} ($CaMg_2$)	E_{st} ($CaAl_2$)
C15	−1.244494	−2.953745
6a	−1.244482	−2.953729
6b	−1.244451	−2.953585
C14	−1.244402	−2.953313

From Tables 1 and 2 it is evident that E_{es} dominates the determination of the stable crystal structures of both $CaMg_2$ and $CaAl_2$. Thus, the theoretically predicted most stable structure in both cases is C15, i.e. the one with the lowest E_{es}. For $CaAl_2$ this is a success, because the theory yields already the correct structure C15. However, for $CaMg_2$, the observed structure C14 comes as the most unstable structure according to theoretical results. Although E_{bs} favours C14 for $CaMg_2$, the differences in E_{bs} are smaller than those in E_{es}, which favours C15.

We assume that the disagreement in case of $CaMg_2$ is due to the influence of the strong charge variations in the IMC $CaMg_2$. The deviations from the mean density may not be taken into account by linear screening.

It is well known that the difference in the electronegativity of the two components of a metallic alloy plays an important role in the determination of its stable crystal structure (HUME-ROTHERY; WABER et al.). One can, to some extent, conclude the occurrence of charge transfer between the two components of an alloy, if they have different electronegativities. This is the case in $CaMg_2$, because Mg has a higher electronegativity (ALLRED) (= 1.31) than Ca (= 1.0). Thus, we believe that a charge transfer takes place from Ca to Mg atoms in $CaMg_2$, not included in linear screening.

We describe the charge transfer in an intermetallic compound AB_2 by a parameter Q defined by $n_B = Q^3 n$. n_B is the charge density on atoms of type B and n is the charge density of the homogeneous electron gas. For a given Q or n_B the charge density n_A results from the neutrality condition:

$$n[V_A + 2V_B] = n_A V_A + 2n_B V_B \tag{8}$$

assuming a close packing of rigid spheres in AB_2 we have $V_A = \frac{4}{3}\pi r_A^3$ and $V_B = \frac{4}{3}\pi r_B^3$, where the radii fulfil the relation: $r_A = \sqrt{1.5}\, r_B$ (SCHULZE). We include higher order effects approximately by screening the A and B cores by n_A and n_B, respectively, instead of n. Through this process the calculated values of the energy wave number characteristics will change depending on the value of Q. In our simple approximation F^{N}_{B-B} is spread by a scale factor Q similating the change of k_F at B places, whereas F^{N}_{A-A} is pressed correspondingly. The $\bar{F}^{N}_{A-B}$ branch is left unchanged. The results of our calculations of E_{st} including charge transfer are given in Table 3 and figures 1 and 2 for $CaMg_2$ and $CaAl_2$. The value $Q = 1.00$ corresponds to calcula-

Table 3
E_{st} (a.u.) of C15, 6a, 6b and C14 calculated with different charge densities. Q^3 is charge density on Mg($CaMg_2$) or Al($CaAl_2$) relative to homogeneous electron gas. $Q^3 = 1$ corresponds to calculations without charge transfer. The most stable structure is underlined for each Q^3 value

Q^3	$-E_{st}$ (a.u.)			
	C15	6a	6b	C14
$CaMg_2$				
1	1.244494	1.244482	1.244451	1.244402
1.015	1.244956	1.244952	1.244923	1.244867
1.030	1.245408	1.245411	1.245384	1.245326
1.046	1.245851	1.245858	1.245837	1.245790
1.061	1.246284	1.246294	1.246285	1.246259
1.077	1.246707	1.246719	1.246723	1.246718
1.093	1.247121	1.247134	1.247150	1.247169
1.109	1.247523	1.247536	1.247561	1.247597
1.125	1.247919	1.247929	1.247958	1.248005
1.158	1.248680	1.248687	1.248722	1.248784
1.191	1.249394	1.249412	1.249454	1.249520
1.225	1.250021	1.250071	1.250130	1.250199
1.260	1.250686	1.250740	1.250798	1.250859
1.331	1.251912	1.251966	1.252020	1.252075
1.424	1.253358	1.253345	1.253356	1.253390
1.521	1.254724	1.254742	1.254773	1.254818
$CaAl_2$				
1	2.953745	2.953729	2.953585	2.953313
1.015	2.956421	2.956399	2.956237	2.955938
1.030	2.959114	2.959033	2.958825	2.958491
1.046	2.961732	2.961601	2.961358	2.961002
1.061	2.964284	2.964097	2.963817	2.963443
1.093	2.969143	2.968863	2.968536	2.968164
1.125	2.973804	2.973460	2.973104	2.972736
1.158	2.978324	2.978000	2.977639	2.977242
1.191	2.982765	2.982477	2.982099	2.981633
1.208	2.985032	2.984730	2.984332	2.983837
1.225	2.987340	2.987021	2.986609	2.986103
1.242	2.989714	2.989378	2.988962	2.988466
1.260	2.992212	2.991849	2.991430	2.990956
1.331	3.003863	3.003371	3.002943	3.002578
1.424	3.017335	3.018499	3.019463	3.020226
1.521	3.044264	3.045176	3.046113	3.047075

tions without charge transfer. Although, in case of $CaAl_2$ the correct structure C15 is already predicted with $Q = 1$, we wish to study the effect of including charge transfer, because the difference in electronegativity of AL (= 1.61) and Ca (= 1) is large.

From Table 3 and Figure 1 we see that the most stable structure of $CaMg_2$ changes according to the amount of charge transfer allowed as follows:

a) In the range $Q^3 = 1$ to 1.024 it is C15,
b) from $Q^3 = 1.025$ to 1.071 it becomes 6a,
c) around $Q^3 = 1.077$ it changes to 6b,
d) finally starting from $Q^3 = 1.081$ up to $Q^3 = 1.521$ it becomes C14.

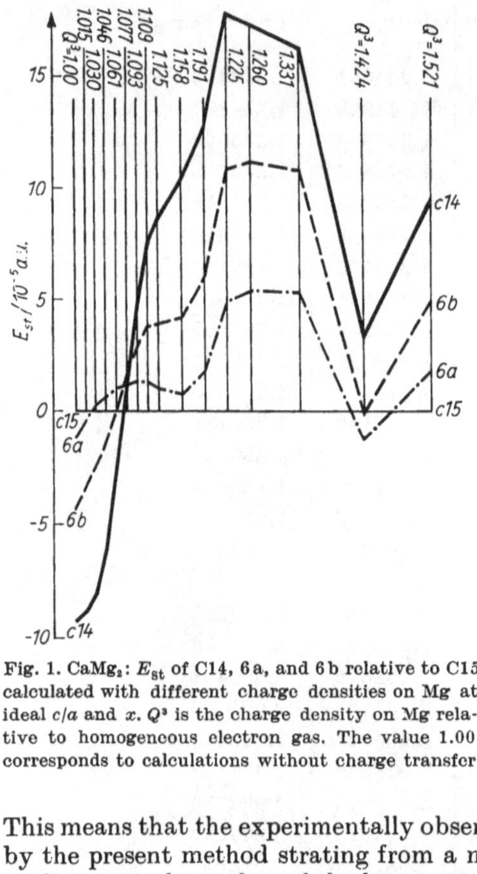

Fig. 1. $CaMg_2$: E_{st} of C14, 6a, and 6b relative to C15 calculated with different charge densities on Mg at ideal c/a and x. Q^3 is the charge density on Mg relative to homogeneous electron gas. The value 1.00 corresponds to calculations without charge transfer

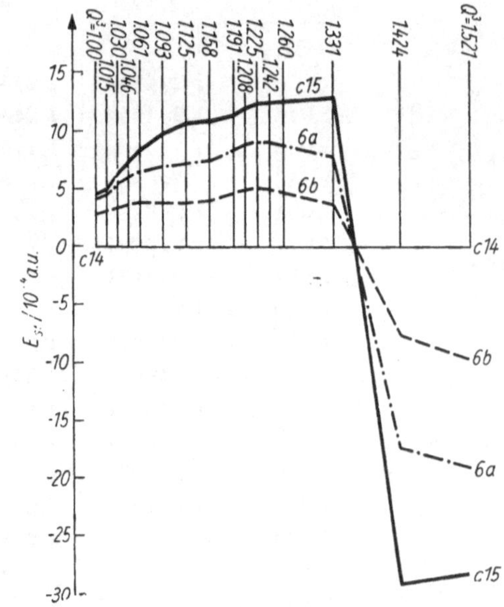

Fig. 2. $CaAl_2$: E_{st} of C15, 6a, and 6b relative to C14 calculated with different charge densities on Al at ideal c/a and x. Q^3 is the charge density on Al relative to homogeneous electron gas

This means that the experimentally observed structure of $CaMg_2$ is correctly predicted by the present method strating from a mean charge density on Mg atoms n_{Mg} which is about 8% above that of the homogeneous electron gas n. This behaviour continues up to the $n_{Mg} = 1.52n$, i.e. n_{Mg} is 52% higher than n. However, the maximum difference $\Delta E_{st} = |E_{st}(C14) - E_{st}(C15)|$ is achieved at $Q^3 = 1.225$, then it drops at $Q^3 = 1.424$ and rises again at $Q^3 = 1.521$.

In case of $CaAl_2$ the theory predicts the observed C15 structure in the whole range $Q^3 = 1$ to 1.36, i.e. up to n_{Al} which is 36% above n. The difference $\Delta E_{st} = |E_{st}(C15) - E_{st}(C14)|$ increases in favour of C15 with ascending value of Q^3, reaching its maximum at $Q^3 = 1.331$. However, if more charge transfer is allowed, C14 structure will become the theoretically most stable one in the range $Q^3 = 1.361$ to 1.521.

We conclude that including charge transfer makes it possible to predict the correct structure for $CaMg_2$. Also C15 remains the most stable structure of $CaAl_2$ over quite a wide range of Q. Thus charge transfer plays an important role in IMC. It may be included approximately by defining new screened pseudoatoms, which result from a screening with individual k_F-values instead of the mean one. With these new screened pseudoatoms second-order perturbation theory leads to the experimentally observed structure. The exact value of Q can be determined by calculating the stacking fault energy of $CaMg_2$ and $CaAl_2$ and comparing it with the experimental values, as we did for $MgZn_2$ (RENNERT, RADWAN). Unfortunately, there are, to our knowledge, no such experimental results for $CaMg_2$ and $CaAl_2$.

References

ALLRED, A. L.: J. Inorg. Nuclear Chem. **17** (1961) 215
APPAPILLAI, M., HEINE, V.: Tech. Rep. No. 5, Solid State Theory Group, Cavendish Laboratory, Cambridge 1972
HEINE, V., WEAIRE, D.: Solid State Physics **24**, New York 1970
HUME-ROTHERY, W. H.: Phase Stability in Metals and Alloys, ed. P. S. Rudman, New York 1967
INGLESFIELD, J. E.: J. Phys. C2 (1969) 1285
MROSAN, E., JOHN, W., ESCHRIG, H.: phys. stat. sol. (b) **51** (1972) 793
PEARSON, B. W.: A Handbook of Lattice Spacings and Structures of Metals and Alloys, Vol. 2, New York 1967
RADWAN, A. M.: Cryst. Res. Technol. **23** (1988) 159
RADWAN(1), A. M.: Cryst. Res. Technol. **23** (1988) 537
RADWAN, A. M., RENNERT, P.: Cryst. Res. Technol. **10** (1975) 865
RADWAN, A. M., SALEH, A. A.: Fizika **19** (1) (1987) 17
RENNERT, P., LEHMANN, G., PAASCH, G., TAUT, M.: phys. stat. sol. **35** (1969) 667
RENNERT, P., RADWAN, A. M.: phys. stat. sol. (b) **79** (1977) 167
SCHULZE, G. E. R.: Metallphysik, Berlin 1967
SHAW, R. W.: Phys. Rev. **174** (1968) 769
SHAW, R. W.: J. Phys. C **2** (1969) 2350
SHAW, R. W.: J. Phys. C 3 (1970) 1140
SINGWI, K. S., SJOELANDER, A., TOSI, M. P., LAND, R. H.: Phys. Rev. **B1** (1970) 1044
TAUT, M., PAASCH, G.: phys. stat. sol. (b) **51** (1972) 295
WABER, J. T., GSCHNEIDER, J. K., LARSON, A., PRINCE, M.: Trans. MS AIME **227** (1963) 717

(Received November 12, 1987;
accepted November 20, 1987)

Author's address:

Dr. A. M. RADWAN
Theoretical Physics Lab., National Research Centre
Tahrir Street
Cairo — Dokki
AR Egypt

Cryst. Res. Technol.	23	1988	5	699–703

A. M. RADWAN

Theoretical Physics Lab., National Research Centre
Cairo, Egypt

Radial Distortions of the Fermi Surfaces of Rubidium and Cesium

The radial distortions of the Fermi surfaces of Rb and Cs are calculated in the three principal directions 100, 110 and 111. The calculations are carried out using different values of the model potential parameters and the effective charge. Our results are compared with available experimental and theoretical ones.

Die radialen Fermiflächenverzerrungen von Rb und Cs werden in den drei Hauptrichtungen 100, 110 und 111 berechnet. Die Rechnungen werden mit verschiedenen Werten der Modellpotentialparameter und der effektiven Ladung durchgeführt. Unsere Ergebnisse werden mit den vorhandenen experimentellen und theoretischen Werten verglichen.

1. Introduction

The shape of the Fermi surface is quite sensitive to the choice of the metal potential and the comparison between the calculated Fermi surface distortions and the observed ones provides an accurate test for the crystal potential.

The de Haas-van Alphen effect was observd for Rb by SCHOENBERG and STILES and for Cs by OKUMURA and TEMPELTON.

Theoretical studies on the Fermi surface of Rb and Cs began with the systematic calculations of HAM using the quantum defect QD method. MAHANTI and DAS have calculated the Fermi surface of Cs by using an OPW procedure with several crystal potentials and examined which potential was best to explain the experiments. LAWRENCE has applied the KKRZ method to obtain Fermi surface distortions of Rb and Cs. The agreement with experimental measurements is good for Rb, but not for Cs where contact with the zone boundary was predicted. Fermi surfaces of Rb and Cs were also evaluated by INOUE et al. using APW method with various crystal potentials. The purpose of their calculation was to examine the accuracy of the ab initio calculation to reproduce the observed Fermi surface and to examine what potential is the best. They suggested that the best potential should include the effects of the crystal. MATSUURA used the Heine-Abarenkov-Animalu type model potential (HAAMP) to calculate Fermi surface distortions of Rb and Cs and obtained better agreement with experiment for Cs than for Rb. MATSUURA et al. applied the energy independent model potential (EIMP) to first principles calculations of Fermi surface distortions for the alkali metals. They found that the EIMP results are better than the existing APW results for Rb and Cs. They also concluded that results obtained by SHAW's opimized model potential OMP are much worse than those calculated by the EIMP for heavier metals. This is mainly due to the uncertainty in the model potential parameters of Rb and Cs. This point is worth investigating.

The purpose of the present paper is to determine the Fermi surface of the heavy alkali metals Rb and Cs using the OMP (SHAW) with different values of the model potential parameters and the effective charge. The results are then compared with the available theoretical and experimental ones to examine various model potential parameters. Of course, there is no guarantee that such a procedure will yield better results. It is, however, an approach that must be investigated.

2. Method of calculation

The secular equation for energy bands has to be solved to determine the Fermi energy E_F. Then we find the wave vector $\boldsymbol{k}$ in each direction which satisfies the relation $E_k = E_F$. The values of $\boldsymbol{k}$ thus determined describe the shape of the Fermi surface. However, there is no simple way of determining the Fermi energy, but we estimate it by the formula (INOUE et al.)

$$E_F = (3E_{100} + 4E_{111} + 6E_{110})/13 \tag{2.1}$$

where E_{100}, E_{111}, and E_{110} are the energy at the free electron Fermi sphere along principal axes. Although this expression is not always accurate enough, the Fermi surface of the alkali metals is practically spherically symmetric, so that this expression is believed to be a good approximation. After determining E_F, we calculate the radial distortions of the Fermi surface δ_{100}, δ_{111}, and δ_{110} defined by

$$\delta_{xyz} = 10^4(k_{xyz} - k_F)/k_F \tag{2.2}$$

where k_{xyz} is the wave-vector length at the Fermi level in the (x, y, z) direction and k_F is the free-electron Fermi-sphere radius. Besides distortion of the Fermi surface along three principal axes, we also calculate the relative distortion between three axes, which is given by the quantities

$$X = \delta_{100} - \delta_{111}\,, \qquad Y = \delta_{110} - \delta_{111} \tag{2.3}$$

X and Y have the advantage of being less dependent on determination of the Fermi energy (INOUE et al.).

3. Results and discussion

The lattice constants (a.u.) used in our calculations are 10.552 (Rb) and 11.424 (Cs) corresponding to $k_F = 0.3694$ (Rb) and $k_F = 0.3412$ (Cs). The model potential parameters are those of APPAPILLAI and HEINE. The exchange and correlation function of SINGWI et al. is chosen together with linear screening.

Special attention has been paid to the convergence of the one-electron energies, owing to the error relation (DAGENS, PERROT)

$$\Delta k = (1/2k_F)\,\Delta E \tag{3.1}$$

which means that an accuracy of 10^{-4} on the distortion, that is on k, requires an identical absolute accuracy on E (since $2k_F \sim 1$). For this reason, we have used an unsymmetrical set of 42 plane waves for Rb and Cs.

It should be mentioned that SHAW's OMP is non-local and energy-dependent. However, to save lot of computation time, the energy-dependence of the OMP is neglected in the present calculations but the non-locality is fully retained.

To investigate the uncertainty in the model potential parameters, we perform several calculations using different values of these parameters. Comparison with experimental Fermi surface distortions can serve as a criterion to find the best set of model potential parameters.

In the present paper, Fermi surface distortions of Rb and Cs are calculated eight times using the following sets of model potential parameters:

a) Those given by APPAPILLAI and HEINE
b) A_2 increased by 10%,
c) A_2 decreased by 10%,
d) changing the sign of dA_2/dE,
e) A_2 doubled,
f) using effective charge $Z = 1 + d$ instead of $Z = 1 - d$ as implied by the theory (APPAPILLAI, HEINE). The reason is that the former value leads to a correct prediction of the observed bcc structure for both Rb and Cs (RADWAN (1)),
g) A_0, A_1, and A_2 all increased by 50%,
h) A_0, A_1, and A_2 all doubled.

Of course, the effective charge had to be recalculated for each of the above cases, except a) and f).

We have solved the 42 × 42 secular determinant numerically for each of the above mentioned cases. Our results for the radial distortions of the Fermi surfaces in Rb and Cs are compared with available theoretical and experimental results in Table 1

Table 1
Various theoretical and experimental values of the radial distortions of the Fermi surfaces for Rb and Cs. Present results are denoted by the letters a) to f). See text for more details

Method	δ_{100}	δ_{111}	δ_{110}	X	Y
Rb					
Expt.	−23	−46	95	42.2*)	154.6*)
EIMP	−24	−78	69	54	148
KKRZ	−50	−40	90	−10	130
APW	−4.9	−83.8	58	78.9	141.8
HAAMP	−2.7	−94.9	71.7	92.2	166.6
a) to e)	−5.9 to −7	−5 to −6.2	6.4 to 7.6	−0.1 to −1	11.4 to 13.7
g)	−42.7	−11.8	31.6	−30.9	43.4
h)	−153	−28	125	−125	153
f)	4.2	−0.3	−1.9	4.5	−1.6
OMP	3.1	−1	−1	4	0
QD	−200	−200	180	0	380
Cs					
Expt.	−94	−143	331	87*)	529*)
EIMP	−71	−158	169	86.5	327
HAAMP	−31	−199	190	168	389
OPW	−83	−295	320	212	615
KKRZ	40	−50	–	90	–
APW	−98.3	−352	515	254	867
a) to e)	−11.9 to −13.6	−9.2 to −10.6	12.3 to 14.2	−2.7 to −3	21.5 to 24.8
g)	−47.7	−13.1	35.6	−34.6	48.7
h)	−170.4	−28.8	142	−141.6	170.8
f)	7	−1	−2.9	8	−1.9
OMP	5	−1.5	−1.6	6.7	0
QD	−600	−600	830	0	1430

*) Analysis by LEE

Analysis of the results in Table 1 shows that:

1. The experimentally observed Fermi surface distortions of Cs are three to four times those of Rb. Consequently, one may conclude that Cs should have much stronger potential than Rb. However, we can see that this is not the case by comparing the model potential parameters of both metals (APPAPILLAI, HEINE). In general, the deviations between theoretical and experimental results of Cs are higher than in case of Rb.
2. The different results obtained by the present OMP case a) and, OMP of MATSUURA et al., though using the same model potential parameters of APPAPILLAI, HEINE arise because: i) we use an unsymmetrical set of plane waves which are closer to the origin of the irreducible part of the Brillouin zone, ii) we neglect the energy dependence of the model potential and iii) different exchange and correlation functions are used.
3. For both Rb and Cs the best theoretical results are those of MATSUURA et al. (EIMP) and the worst are those of HAM (QD). However, we think that there should be an error in HAM's interpolation procedure for the calculation of Fermi surface distortions because his band structure results of Rb and Cs are much better.
4. The present results a) to e) are very small compared to experiment leading to a far less distorted Fermi surface than the actual one. However, δ_{100} has the second best fitting to experimental value for Rb after EIMP. We see also that change of A_2 with -10% to 100%, or reversing the sign of dA_2/dE has only a very small influence on the results, which can be practically neglected.
5. In case f), taking $Z = 1 + d$ worsens the results specially in the directions 100 and 110 for both Rb and Cs contrary to previous results obtained for K (RADWAN). Nevertheless, the X values are improved to come at third place after EIMP and APW for Rb and EIMP and KKRZ for Cs.
6. In calculation g), 50% increase of A_0, A_1, and A_2 increases the distortions considerably in the directions 100 and 110 but not as much in the 111 direction. This tendency is maintained when the potential parameters are further increased by 100%, case h), leading to an overestimated δ_{100} (much higher than observed value) and underestimated δ_{111} for both Rb and Cs. On the other hand the value of Y obtained in case h) for Rb is almost identical with experiment and δ_{111} comes in second place after the KKRZ result.
7. We note that no single set of results has best fit to all of the measured values. However, if we take the sum of the deviations between calculated and observed distortions (in percent %) as a criterion to examine and judge the different theoretical results we find that: In case of Rb the order of the results according to best fitting with experiment is: EIMP, KKRZ, APW, HAAMP, a), g), OMP, f), h) and QD respectively. For Cs it becomes: EIMP, HAAMP, OPW, g), APW, a), OMP, f), h) and QD, respectively.

We conclude that Fermi surface distortions of Rb and Cs calculated using OMP method with parameters of APPAPILLAI and HEINE have only qualitative but no quantitative agreement with experiment. Thus OMP is not so successful for heavy alkali metals as it was for the light ones (RADWAN). Increasing the potential parameters in case of Rb may improve some single values but not the results as a whole. For Cs an increase of all potential parameters by 50%, case g), yields OMP results which are closer to experiment than APW values. Present OMP calculations underestimate δ_{111} and δ_{110}, even if the potential is raised by 100% (50% for δ_{110} of Rb). On the other hand, δ_{100} is overestimated by an increase of 50% for Rb and 100% for Cs. This feature of OMP results is completely reversed in EIMP, APW, HAAMP methods where δ_{111} is overestimated for both Rb and Cs. We do not agree with the conclusion of LAWRENCE predicting that the Fermi surface of Cs touches the N face of the Brillouin zone. As noted by INOUE et al., the Fermi surfaces in Rb and Cs are rather sensitive to the position of the d-levels above the Fermi levels. Therefore, the fact that EIMP yields reasonable results in Rb and Cs seems to indicate that EIMP reasonably describes the effects of the d-levels above the Fermi level than OMP do.

Finally, we think that preserving the energy-dependence of the OMP and not neglecting it as in present calculation, may lead to better Fermi surface distortion results for beavier metals like Rb and Cs.

References

APPAPILLAI, M., HEINE, V.: Tech. Rep. No. 5, Solid State Theory Group, Cavendish Laboratory, Cambridge 1972
DAGENS, L., PERROT, F.: Phys. Rev. B 8 (1973) 1281
HAM, F. S.: Phys. Rev. **128** (1962) 2524
INOUE, S. T., ASANO, S., YAMASHITA, J.: J. Phys. Soc. Japan **30** (1971) 1546
LAWRENCE, M. J.: J. Phys. **F1** (1971) 836
LEE, M. J. G.: Phys. Rev. **178** (1969) 953
MAHANTI, S. D., DAS, T. P.: Phys. Rev. **183** (1969) 674
MATSUURA, M.: Phys. Lett. **50A** (1974) 131
MATSUURA, M., WOO, C. H., WANG, S.: J. Phys. **F5** (1975) 1849
OKUMURA, K., TEMPELTON, I. M.: Proc. Roy. Soc. **A287** (1965) 89
RADWAN, A. M.: to be published in FIZIKA **19** (3) (1987)
RADWAN(1), A. M.: Cryst. Res. Technol. **23** (1988) 159
SHAW, R. W.: Phys. Rev. **174** (1968) 769
SHOENBERG, D., STILES, P. J.: Proc. Roy. Soc. **A281** (1964) 62
SINGWI, K. S., SJOELANDER, A., TOSI, M. P., LAND, R. H.: Phys. Rev. **B1** (1970) 1044

(Received November 16, 1987;
accepted November 23, 1987)

Author's address:

Dr. Ahmed M. RADWAN
Theoretical Physics Laboratory
National Research Centre
Tahrir Street
Cairo-Dokki, A. R. Egypt

Cryst. Res. Technol.	23	1988	5	705–712

R. Handke, G. Lippert

Akademie der Wissenschaften der DDR, Institut für Halbleiterphysik, Frankfurt/Oder

Characterization of Plasma Etching of Molybdenum Polycid Conductor Stacks in Cl_2/CF_4 and $Cl_2/CF_4/O_2$

The present paper is concerned with an extremly anisotropic two-step patterning process for Mo2Si/poly-Si double layers with an etch-stop on thick P-doped CVD-SiO_2. As etching gas $Cl_2/CF_4/O_2$ and Cl_2/CF_4 are used. By means of etch rate dependencies, volatilities of possible reaction products and results, obtained from the literature it is attempted to develop a simple qualitative model on the chemism of the process.

Der Artikel beschreibt einen äußerst anisotropen plasmachemischen Zwei-Phasen-Strukturierungsprozeß für MoSi/poly-Si-Doppelschichten mit Ätzstopp auf dickem P-dotiertem CVD-SiO_2. Als Ätzgase werden Cl_2/CF_4 und $Cl_2/CF_4/O_2$ verwendet. Anhand von Ätzratenabhängigkeiten, Flüchtigkeiten möglicher Reaktionsprodukte und Ergebnissen aus der Literatur wird versucht, ein einfaches qualitatives Modell zum Chemismus des Prozesses zu entwickeln.

1. Introduction

Highly doped poly-Si is a contacting material, so far most frequently used for the manufacture of MOS-circuits. But as a result of its high resistivity and the increasing reduction of its pattern dimension it becomes more and more a power-limited factor.

Polycides (poly-Si/silicide-double layers), silicides (e.g. $MoSi_2$, $TaSi_2$, $TiSi_2$) or also metal single layers (e.g. W, Mo) constitute an alternative (Murarka). The small pattern dimensions (1 μm) common in today's VLSI-technology require the application of dry etching technologies.

When applying dry or plasma etching processes, we are faced with the difficulty of simultaneously meeting numerous requirements, such as anisotropy of the process, selectivity with regard to base and mask material, uniformity of the process across wafer and batch, freedom from impurities and radiation damages etc.

The etching of polycide stacks is considered to be particularly complicated, since the above requirements (preferably those of anisotropy) must be fulfilled for both layer materials at the same time. Different etching behaviours of the two layers often give rise to discontinuities in the edge profile (Bennett, Chow).

Moreover, difficultly etchable intermediate layers (interface Poly-Si/silicide) frequently make the problem more intricate (Whitecomb, Jones). The papers of Chow and Steckl and Peccoud and Lapporte give a good survey of the large number of papers on the plasma chemical patterning of silicide-containing layer systems. What strikes most is the large number of the applied gases, with chlorine-containing and/or gas mixtures and high frequencies (e.g. 13.56 MHz) prevailing in a RIE-equipment.

The present work is concerned with experiments extended to the plasma-chemical patterning of Mo2Si/poly-Si-double layers with Cl_2/CF_4- and/or $Cl_2/CF_4/O_2$-gas mixtures, thereby endavouring to develop simple qualitative model concepts on the chemism of the process.

2. Experimental work

The experiments were carried through on a single wafer etcher (diode system, Fig. 1), the most important parameters of which are:

- 27 MHz-plasma exitation frequency, which may alternatively be supplied to the substrate (RIE-mode) and/or non-substrate electrode (PE-mode)
- electrode diameter 125 mm
  electrode spacing 25 mm
  electrode material Al_2O_3
- automatic equipment ellipsometer for the 'in situ' etch rate or end point determination.

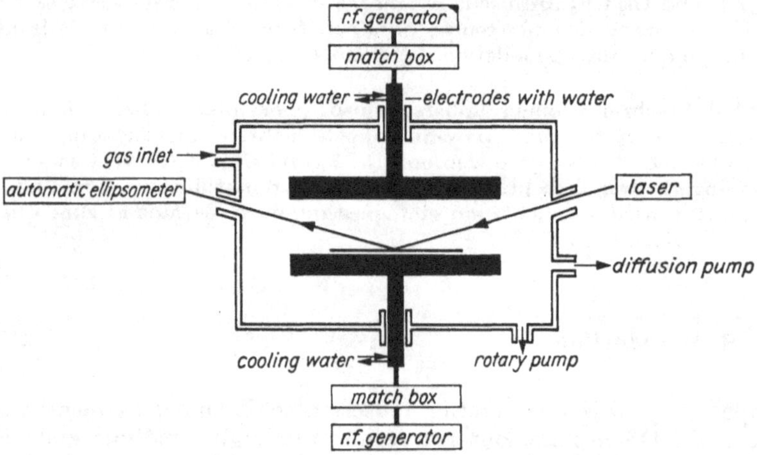

Fig. 1. Diagram of the diode system

The vacuum system of the etcher has an oil diffusion pump for cleaning purposes, producing a pressure of 6.6×10^{-3} Pa in the recipient before etching. A rotary pump serves as an operating pump. Cl_2/CF_4 (ratio 1 :1) and/or Cl_2/CF_4 (ratio 1 :1)with an O_2-addition are used as etching mixtures.

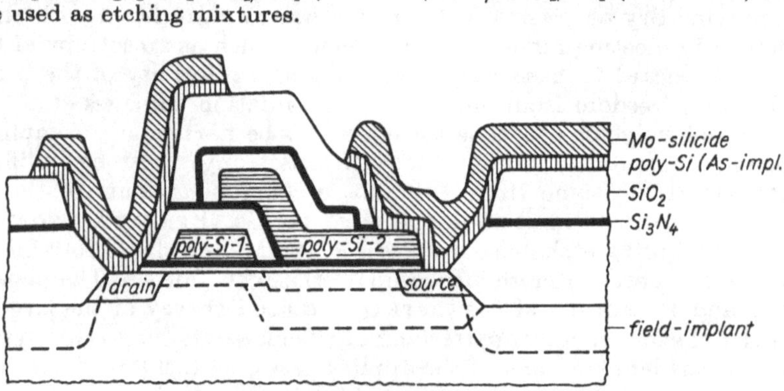

Fig. 2. Schematic drawing of cross-sectional view of a storage cell

The operating pressure was varied between 2.6 Pa and 53 Pa, the power between 70 W and 110 W.

The applied molybdenum silicide was deposited in accordance with the mosaic target methode (LIPPERT et al.) and generally silicided after etching, owing to the problems, referred to in literature. The layer thickness is 270 nm.

LP-poly-Si, 200 nm thick, being implanted with 1×10^{-16} arsenic At/cm^2, served as a base in the polycide stack. Annealing was also performed only after etching.

The etch stop was realized on a thick phosphorus-doped CVD-SiO_2-layer.

3. Results

Comprehensive investigations with Cl_2/CF_4 (1:1) in the given parameter field show that a complete patterning of Mo2Si/poly-Si-double layers is not possible.

Figure 3 shows a polycide line of 2 micron with photoresist mask after try to pattern with Cl_2/CF_4. Residual stacks can clearly be seen between the lines.

No improvements are achieved by changing the parameters pressure, power, and time. Microprobe analyses showed that difficultly volatile Mo-containing components, presumbly chlorides, prevent a full patterning of the stack layer, whereas patterning of poly-Si single layer is possible at the same operating point.

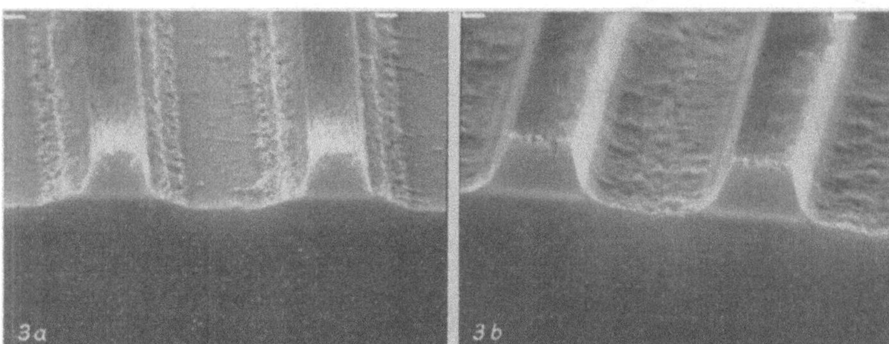

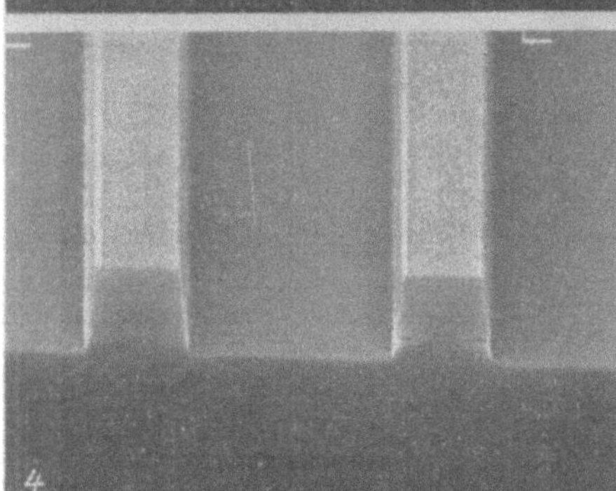

Fig. 3. SEM photos of 2 μm polycide lines with photoresist mask etched using Cl_2/CF_4 (1:1) a) Etching parameters: RIE, p = 8 Pa, P = 100 W, U_B = −230 V, t_E = 13 min; b) Ething parameters: p = 8 Pa, P = 50 W, 1. step: RIE, U_B = −130 V, t_E = 30 min; 2. step: PE, U_B = −90 V, t_E = 20 min

Fig. 4. SEM photos of 2 μm polycide lines with photoresist mask (undoped and As implanted line). Etching parameters: RIE, p = 8 Pa, P = 50 W, U_B = −130 V, t_E = 8.9 min)

The profile section within the poly-Si-layer near the mask correlates well with the penetration depth of the As-implant (60−80 nm) and may possibly be attributed to a considerably increased lateral etching rate in this region.

From Table 1 it is seen that silicon can enter into sufficiently volatile combinations with fluorine (SiF_4), but also with chlorine ($SiCl_4$).

Against this molybdenum has a well volatile fluoride (MoF_4), but no easily volatile chloride!

Comparing the volatilities of oxychlorides and/or oxyfluorides to those of the chlorides and/or fluorides of molybdenum, it is seen that the well-known possible oxyfluorides are far worse volatile than the fluorides, volatile Mo-oxychlorides being far more easily possible than Mo-chlorides. The fact that molybdenum oxychlorides appear more volatile than the oxyfluorides, is of special interest. The higher volatility of molybdenum oxychlorides compared to Mo-chlorides induced us to carry on the experiments with Cl_2/CF_4 (1:1), while adding O_2.

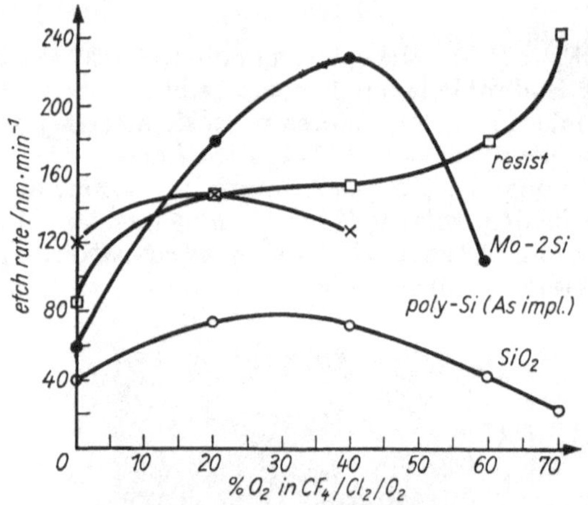

Fig. 5. The etch rates of Mo2Si, photoresist (AZ 1350 H), poly-Si and SiO_2 vs % O_2 in $Cl_2/CF_4/O_2$. Ething parameters: RIE, $p = 8$ Pa, $P = 100$ W, x% O_2 in Cl_2/CF_4 (1:1)

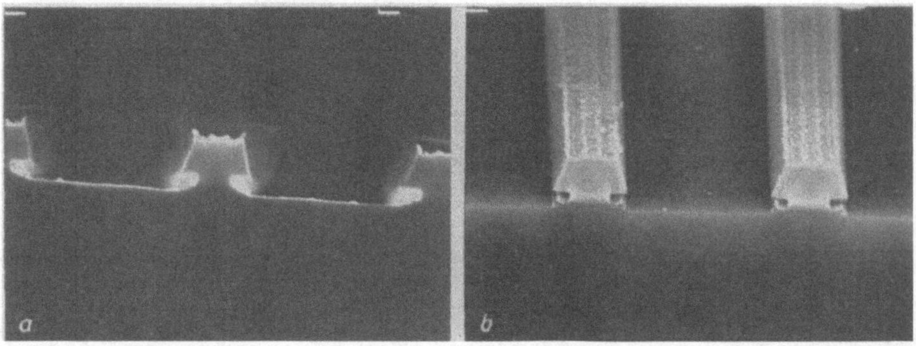

Fig. 6. SEM photos of 2 µm polycide lines with photoresist mask etched using Cl_2/CF_4 and 20% O_2. Etching parameters: RIE, $p = 8$ Pa, $P = 100$ W, $U_B = -270$ V, $t_E = 2.8$ min. a) tilt angle 90°, b) tilt angle 70°

Figure 5 illustrates the dependence of the etching rate on the added O_2-component for the materials under investigation. The Mo2Si-etching rate exhibits a sensitive reaction on the added O_2-component, thereby rising while adding oxygen up to 40%, but beyond that abruptly decreasing. At approximately 40% O_2 the rate achieved is nearly 3 times higher than for a pure Cl_2/CF_4 mixture.

The photoresist etching rate shows a similarly sensitive reaction on the O_2-addition, thereby rising with increasing O_2-fraction. This effect is well-known and serves to 'ash' the photoresist.

The etching rates of poly-Si and SiO_2 (here thermally grown) less sensitivly react on O_2-addition. The selectivity of poly-Si to SiO_2, however, reaches only values of 3. From the standpoint of the choice of an operating point for etching the stack of layers, an intersection of the curves of molybdenum silicide and poly-Si (the same etching rates) appears particularly favourable. Such an intersection exists with an O_2-addition of approximately 10—20%.

Etching experiments on stacks of polycides with Cl_2/CF_4 and O_2 (20%) show that with these gaseous mixtures a full patterning of the double layer free from residues is possible. As compared to mask and underlying poly-Si it is seen, however, that the Mo2Si-layer is perceptibly undercut.

Experiments on Mo2Si-single layers show that Mo2Si can be patterned in an anisotropic way and without an edge displacement in a very limitedly etched process with $Cl_2/CF_4/O_2$, the edge displacement rapidly increasing with potential overetching.

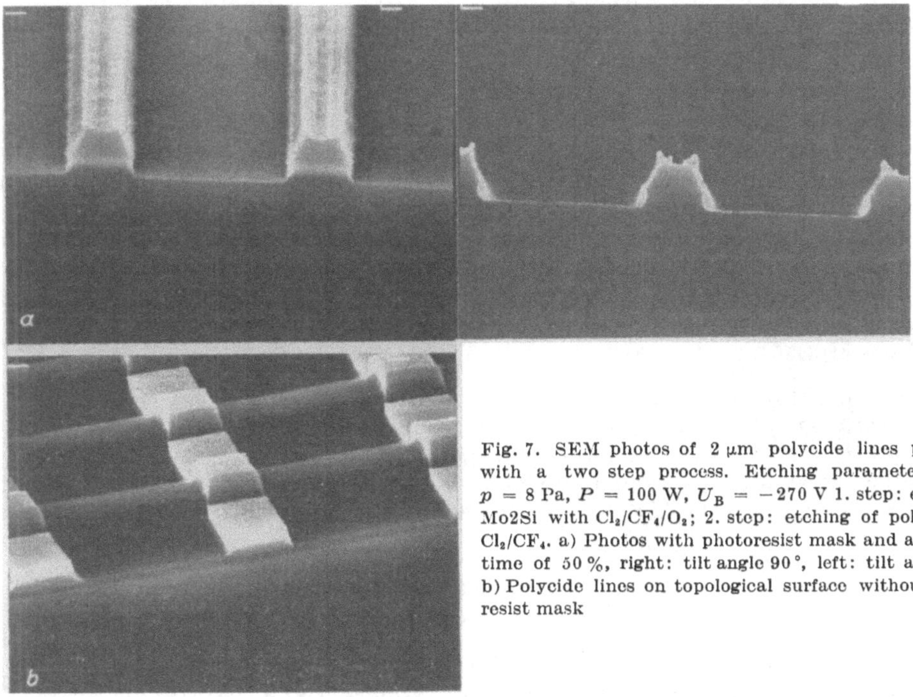

Fig. 7. SEM photos of 2 µm polycide lines patterned with a two step process. Etching parameters: RIE, $p = 8$ Pa, $P = 100$ W, $U_B = -270$ V 1. step: etching of Mo2Si with $Cl_2/CF_4/O_2$; 2. step: etching of poly-Si with Cl_2/CF_4. a) Photos with photoresist mask and a overetch time of 50 %, right: tilt angle 90°, left: tilt angle 70°; b) Polycide lines on topological surface without photoresist mask

From this it is concluded that the etching process should be split into a two step process. Figure 7 shows a 2 micron polycide line with photoresist mask, patterned in a two-step process:

— first step etching of Mo_2Si with $Cl_2/CF_4/O_2$
— second step etching of poly-Si with Cl_2/CF_4.

The final point of the individual steps was determined by means of the automatic ellipsometer. The slight edge displacement, remaining stable even with extremely long overetch times (50%), is clearly seen. This again is of special interest for etching on topological surfaces, when removing potentially remaining socalled 'conductor filaments'.

Table 1
Volatilities of some molybdenum and/or silicon halogenides

Chloride/ oxychloride	boiling point (°C)	fluoride/ oxyfluoride	boiling point (°C)
$SiCl_4$	57.6	SiF_4	−95.5 (subl.)
$SiCl_2$	145.0	Si_2F_6	−18.5
		MoF_6	35.0
$MoCl_5$	268.0	MoF_5	213.6
$MoCl_2$	166.8	MoF_4, MoF_3	non vol.
$MoOCl_4$	subl.	$MoOF_4$	180.0
$MoOCl_3$	100.0 (subl.)		
MoO_2Cl_2	198.9 (subl.)	MoO_2F_2	270.0 (subl.)

4. Discussion

In such a plasma, used for reactive ion etching, two competing processes can take place in parallel with each other-layer deposition and etching. Depending on the parameters selected, the one or the other component may prevail.

In order to achieve an anisotropic etching process, it is essential that a passivation of the sidewalls is achieved on one hand and an etching in vertical direction takes place on the other, with the influence of ions playing a specific role.

For purpose of forming sidewall-passivating layers the following four possibilities derive from MIETH and BARKER:

1. Formation of layers similar to polymeres (e.g. when etching with CHF_3)
2. Chemical passivation by monolayers (e.g. O and/or Cl-adsorption)
3. Redeposition of difficulty etchable material (e.g. SiO_2)
4. Reaction of etching gas and layer material to be etched on relatively difficult volatile reaction products (e.g. $SiCl_4$), which prove to be volatile when exposed to ions only.

The formation of anisotropic etch profiles, however, requires the influence of ions, with two general possibilities of causing anisotropic etching by ion bombardment:

1. The reactants cannot spontaneously etch the layer to be etched.
Ion influence: causing and/or supporting the formation and desorption of reaction products;
2. The reactants can spontaneously etch the layer, but there are particles retarding the etching process.
Ion influence: causing the desorption of reaction products.

In both cases a removal by etching is observed merely in places directly exposed to the ionbombardment, giving rise to an anisotropic etching process.

With the aid of our results, the above mentioned volatilities of possible reaction products and the results obtained from literature, we endavoured to gain simple qualitative model conceptions by means of possible reaction equations of the individual steps:

1) Mo2Si-etching with $Cl_2/CF_4/O_2$

$$CF_4 \rightleftharpoons CF_3 + F\,; \qquad Cl_2 \rightleftharpoons 2\,Cl\,; \qquad O_2 \rightleftharpoons 2O \tag{1.1}$$

$$Si + 4\,F \rightleftharpoons SiF_4 \tag{1.2}$$

$$Mo + O_x + Cl_x \rightleftharpoons MoO_xCl_x \quad (\text{e.g. } MoOCl_3) \tag{1.3}$$

2) poly-Si etching with Cl_2/CF_4

$$CF_4 \rightleftharpoons CF_3 + F\,; \qquad Cl_2 \rightleftharpoons 2\,Cl\,; \qquad O_2 \rightleftharpoons 2\,O \tag{2.1}$$

$$Si + 4\,F \rightleftharpoons SiF_4 \tag{2.2}$$

$$Si + 4\,Cl \rightleftharpoons SiCl_4 \tag{2.3}$$

$$Mo + Cl_x \rightleftharpoons MoCl_x \quad (\text{e.g. } MoCl_5)\,. \tag{2.4}$$

The equations (1.1) and (2.1) suggest the formation of reactive particles (ions and radicals) in the plasma. From numerous publications it is well known and mass- or emission-spectroscopic analyses supplied sufficient proof to the effect that Si can be converted to silicon tetrafluoride in the presence of fluorine (1.2 and 2.2).

While adding oxygen to Cl_2/CF_4-mixture Mo2Si is etched free from residues at an increased etching rate, from which it is concluded that relatively highly volatile Mo-oxychlorides be formed (equ. 1.3). The volatility of the oxychlorides, however, is not so high that a spontaneous etching (isotropic etching process) takes place. Assuming the desorption of the reaction products to be supported by ions a sidewall passivation is achieved on one hand, and a vertical on the other, i.e. anisotropic etching. The equations (2.1) and (2.2) correspond to step 1. The reaction mentioned under (2.3) characterizes the possible reaction of silicon with chlorine.

From literature we know that this process can take place in an anisotropic way, which can be caused either by an Cl-adsorption on the sidewalls or by a reaction of the sidewalls with chlorine to $SiCl_4$ on one hand and an ion-supported vertical etching on the other. While etching poly-Si with Cl_2/CF_4 the edge displacement of the Mo2Si-layer remains constant, from which it is concluded that a sidewall passivation of the Mo2Si-layer be caused by difficulty volatile Mo-chlorides (equ. 2.4).

5. Summary

The plasma etching of Mo2Si/poly-Si double layers is investigated in a single wafer etcher, with Cl_2/CF_4 and $Cl_2/CF_4/O_2$-gas mixtures serving as etch gases.

A full patterning of the double layer in a Cl_2/CF_4 (ratio 1:1)-plasma is seen to be impossible. Microanalyses showed that difficult volatile Mo-compounds (e.g. chlorides) are blocking the etch attack.

A full anisotropic two-step etch process of Mo-polycid is developed, with the Mo2Si etching in $Cl_2/CF_4/O_2$ as the first step and the poly-Si etching in Cl_2/CF_4 as the second step. With the first step it is also possible to etch Mo2Si-single layers in an anisotropic way.

By means of our results and results obtained from the literature, it is attempted to develop simple qualitative model concepts on the chemism of the process.

References

Bennett, R. S., Chow, T. P.: Solid State Technology, April (1985) 193

Chow, T. P., Steckl, A. J.: Plasma Processing, Proceedings of the 4-th Symposium 1983, Vol. 82–100, p. 362–381

LIPPERT, G., PROCOP, M., BORCHARDT, W., SPIESS, L., URWANK, P.: Thin Solid Films **143** (1987) 211
MIETH, M., BARKER, R. A.,: J. Vac. Sci. and Techn. Apr.—Ju (1983) 629
MURARKA, S. P.,: Annular Rev. Mater. Sci. **13** (1983) 117
PECCOUD, L., LAPORTE, PH.: Thin Solid Films **140** (1986) 137
WHITECOMB, E. C., JONES, A. W.: Solid State Technology, April (1982) 121

(Received November 16, 1987;
accepted November 23, 1987)

Authors' address:

R. HANDKE, G. LIPPERT
Institut für Halbleiterphysik der
Akademie der Wissenschaften der DDR
Walter-Korsing-Straße 2, PSF 409
Frankfurt/Oder
DDR-1200

Cryst. Res. Technol.	23	1988	5	K81–K83

Short Notes

R. SRNÁNEK, A. ŠATKA, K. SEIDEL

Microelectronics Department, Slovak Technical University, Bratislava, Czechoslovakia, and Karl-Marx-Universität, Sektion Chemie, Leipzig, DDR

The $K_3[Fe(CN)_6]$–KOH–H_2O Photoetchant for InP

1. Introduction

The importance of InP and (Ga, In) (As, P) materials in optoelectronics is well established. It is known that a poor quality substrate is a major source of imperfections in epitaxial layers. Chemical etching is a simple and universal method for revealing of lattice defects on the single crystals' surfaces and epitaxial layers. Several etchants have been described for (001) InP.

The photoetching method is a very sensitive one for revealing of different defects. It produces also etch features when the etch rate is small (GOTTSCHALCH et al.; WEYHER, GILING). LOURENCO described a $K_3[Fe(CN)_6]$-KOH–H_2O photoetchant with an etch rate of 1.5 $\mu m\ h^{-1}$ for defect characterisation of n-(Ga, In) (As, P).

We report investigations to the revealing of defects in (001) InP substrates and epitaxial layers using an aqueous solution of $K_3[Fe(CN)_6]$ and KOH. The comparison of photoetching, conventional chemical etching with CHU-etchant (CHU, JODLAUK) and the cathodoluminescence mode (CL) of SEM makes it possible to identify kind and position of several defects.

2. Experimental

The (001) InP substrates used here were Sn-doped ($2 \times 10^{18}\ cm^{-3}$) and grown by LEC method. The InP epitaxial layers were undoped and part of an InP/(Ga, In) (As, P)/InP double heterostructure grown by LPE. The photoetchant was mixed from a solution of 15 g $K_3[Fe(CN)_6]$ in 100 ml H_2O and a solution of 15 g KOH in 100 ml H_2O in volume ratio 1:1 before etching. This composition was used without illumination for defect revealing in (001)-(Ga, In) (As, P) epitaxial layers in our earlier work (SRNÁNEK et al.). The etching took place under illumination by a 50 W microscope halogen lamp in the solution stirred by a small blade wheel. The maximum etching times were 15 min. Under these conditions the etch rate was 0.14 $\mu m\ min^{-1}$ for LEC-InP and 0.3 $\mu m\ min^{-1}$ for LPE-InP. The etch features were correlated with defects by additional etching for 10 s with 3 HBr:1 HNO_3. Its correlation with defects is given by CHU and JODLAUK. Prior to photoetching the samples were analysed by integral CL in the SEM TESLA BS-300 (accelerating voltage 30 kV).

3. Results and discussion

Typical etch features produced by the photoetchant on LEC-InP are shown in Figure 1b. From the comparison with Figures 1a and 1c follows that they are caused by inclined dislocations. The feature denoted with S indicates the line segment of a dislocation nearly parallel to the surfaces which is to be seen as a shallow pit after CHU-etching.

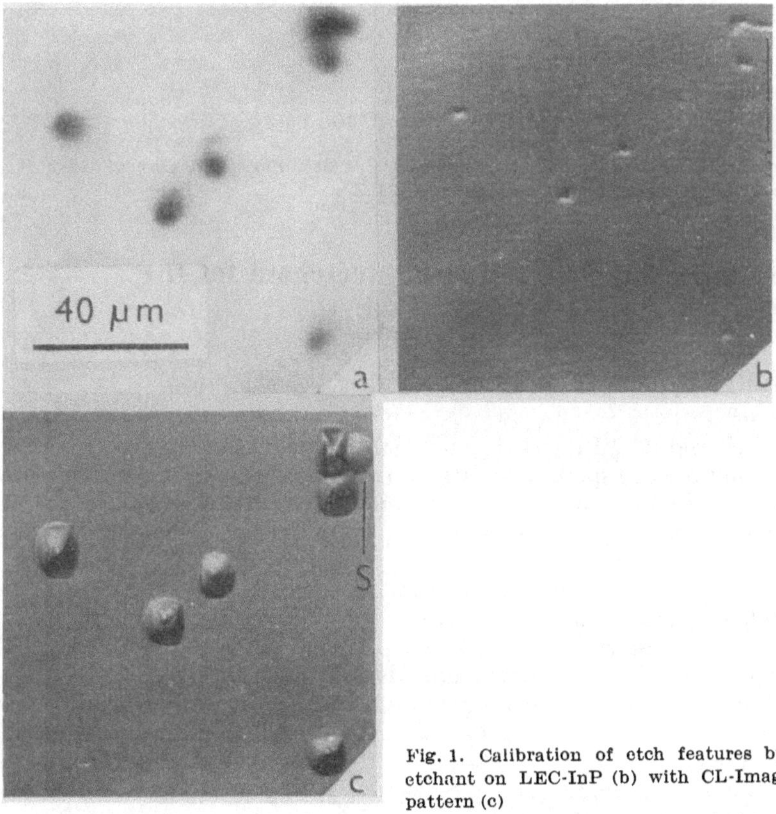

Fig. 1. Calibration of etch features by the described photoetchant on LEC-InP (b) with CL-Image (a) and CHU-etch pit pattern (c)

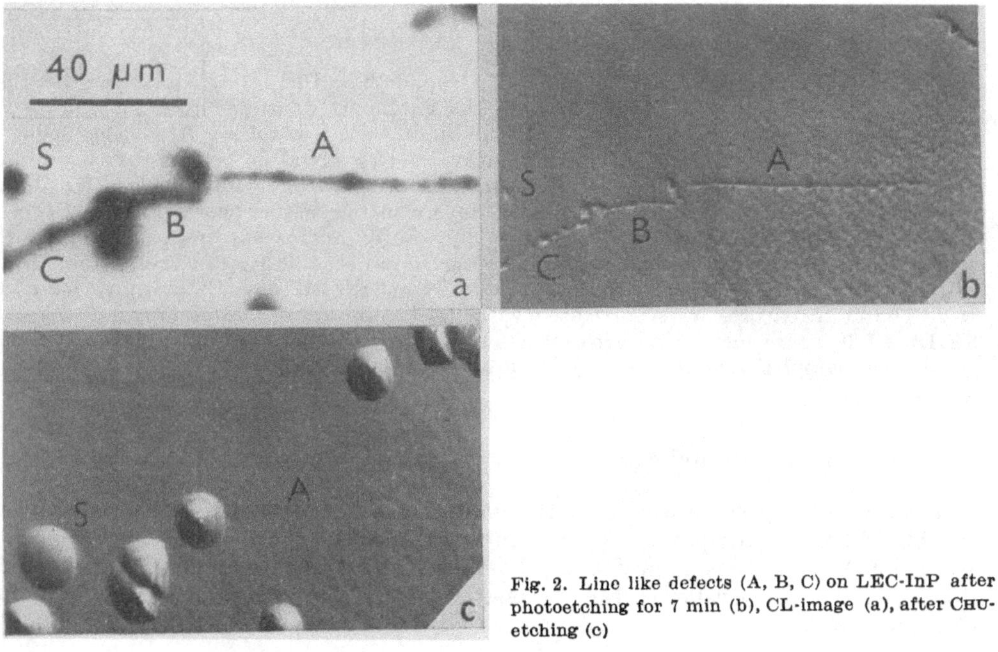

Fig. 2. Line like defects (A, B, C) on LEC-InP after photoetching for 7 min (b), CL-image (a), after CHU-etching (c)

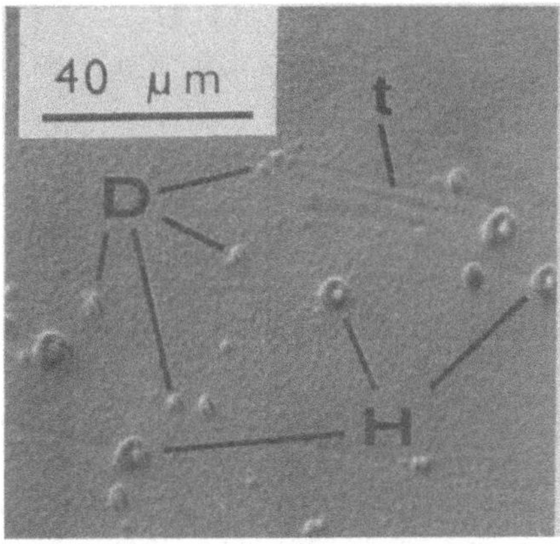

Fig. 3. Photoetched InP epitaxial layer (6 min)
D – etch features, t – "tadpoules",
H – "head of tadpoules"

Fig. 4. Misfit dislocations in InP epitaxial layer photoetched from the "bottom" side

The line features in Figure 2b may be caused by microdefects along a dislocation line parallel to the surface (WAGNER, GOTTSCHALCH; WEYHER, GILING). They disappear after CHU-etching. The feature denoted with S may be caused by a microdefect. Etching of InP epitaxial layers gave patterns (Fig. 3) corresponding to those obtained by other authors with photoetching using the $HF–CrO_3–H_2O$ system (WEYHER, GILING). The surface became rough and etch features are only weakly to be seen. After material-selective etching of InP substrate and (Ga, In) (As, P) layer from the "bottom" side and finally photoetching of the InP layer a better quality surface can be obtained (Fig. 4). Material-selective etching of (Ga, In) (As, P) was done with the solution for photoetching without any light.

Finally a considerable advantage of the described photoetchant is clearly visible. The same composition reveals defects in (001)-InP under illumination and in (001)-(Ga, In) (As, P) without any light (SRNÁNEK et al.) and can be used as a material-selective etchant for (Ga, In) (As, P) on InP. So a heterostructure (001)-InP/(Ga, In) (As, P)/InP can be examined by progressive etching using only the described solution.

References

CHU, S. G., JODLAUK, C. M.: J. Electrochem. Soc. **129** (1982) 352
GOTTSCHALCH, V., SRNÁNEK, R., WAGNER, G.: Mat. Sci. Lett. **1** (1982) 358
LOURENCO, J. A.: J. Electrochem. Soc. **131** (1984) 1914
SRNÁNEK, R., NANDRASKÁ, S., KOVÁČ, J.: Mat. Sci. Lett. **6** (1987) 187
WAGNER, G., GOTTSCHALCH, V.: Cryst. Res. Technol. (Kristall und Technik) **21** (1986) 881
WEYHER, J. L., GILING, L. J.: Appl. Physics **58** (1985) 219

(Received, accepted March 12, 1988)

Authors' addresses:

Rudolf SRNÁNEK, Alexander ŠATKA
Microelectronics Department, Slovak Technical University
Mlynska dolina,
Bratislava
812 19 Czechoslovakia

Kerstin SEIDEL
Sektion Chemie, Karl-Marx-Universität
Talstaße 35
Lripzig 7010, DDR

Cryst. Res. Technol.	23	1988	5	K84–K87

H. Kühne

Akademie der Wissenschaften der DDR
Institut für Halbleiterphysik, Frankfurt (Oder)

A Suplementary Consideration of Lateral Autodoping

Introduction

Lateral Autodoping of epitaxial silicon layers is a consequence of vertical autodoping which causes a parasitic dopant partial pressure not only above buried layer but also out of it. It was suggested earlier that this parasitic pressure in its momentary amount, should be determined by the action of two oppositely directed gas flows of dopants, the flow directed into the gas stream through the reaction chamber, where dopants are lost for parasitic incorporation effects, and that one directed to the wafer surface leading to reincorporation of dopants into the growing epitaxial layer (Kühne). In order to keep gas phase autodoping at a low level the updiffusion into the gas stream should be increased and, in consequence, the finite dopant reservoir in the gas right off the surface rapidly consumed. In a later paper (Kühne) the parasitic dopant pressure has been linked to the dopant adsorbate concentration on buried layer regions, but in this context the possibility has been missed of introducing the quotient formed by dividing dopant flow lost into the gas (I_{gas}) by the readsorption flow (I_{ad}), as a main factor of autodoping control. Recently it has been shown (Kühne) how important the act of neglecting readsorption of dopants will be for describing vertical and also lateral autodoping profiles by the same exponential expression as that applied for intentionally non-steady state doping of epitaxial layers. This expression has been derived by Reif from non-steady state doping experiments and applied to lateral autodoping by Wong.

It is the concern of the present note to demonstrate the fact that readsorption of dopants must not be neglected at all, for it takes part in controling the degree of lateral autodoping.

The parasitic dopant partial pressure

On the background of a theoretical redistribution equilibrium which distributes dopant atoms not only from adsorption sites on the surface to lattice sites of the growing layer but also vice versa, desorption of dopants from buried layer into the gas necessarily must happen. The equations below demonstrate the redistribution equilibrium using arsenic as an example.

$$* + \mathrm{As}_{\mathrm{Si}} \rightarrow \mathrm{As}\,* , \tag{1a}$$

$$N_{\mathrm{As}*} = \frac{(N_{\mathrm{As}_{\mathrm{Si}}})_{T_{\mathrm{epi}}} \cdot N_0}{K_{\mathrm{inc}}} = \frac{(N_{\mathrm{As}^+_{\mathrm{Si}}})_{\mathrm{tot}}}{\left(\dfrac{K_{\mathrm{inc}}}{F_1 \cdot N_0}\right)} \tag{1b}$$

where

$(N_{As_{Si}^+})_{tot}$ at room temperature electrical active total amount of dopants
$(N_{As_{Si}})_{T_{epi}}$ at deposition temperature electrical inactiv amount of dopants
N_0 total amount af adsorption sites

$F_1 = (N_{As_{Si}})_{T_{epi}}/(N_{As_{Si}^+})_{tot}$

Desorption is counterbalanced by diffusion of dopants through the gas and readsorption of dopants on the surface.

$$I_{gas} = I_{des} - I_{ad} \tag{2}$$

where

$$I_{des} = k_{des} \cdot p_{H_2} \cdot N_{As*}\,; \quad As* + H_2 \rightarrow AsH_2 + * \tag{3}$$

$$I_{ad} = k_{ad} \cdot N_0 \cdot p^s_{AsH_2}\,; \quad AsH_2 + * \rightarrow As* + H_2 \tag{4}$$

$$I_{gas} = \frac{D_{gas}}{\delta \cdot RT} \cdot p^s_{AsH_2} \tag{5}$$

The above given equations represent conditions characterized by the application of hydrogen as a carrier gas near to atmospheric pressure and the acceptance of thermodynamic equilibrium composition within the gas above the surface. From the balancing equations we find the parasitic dopant pressure $p^s_{AsH_2}$ to be determined by the following expression:

$$p^s_{AsH_2} = \frac{k_{des} \cdot p_{H_2} \cdot N_{As*}}{\left(k_{ad} \cdot N_0 + \dfrac{D_{gas}}{\delta \cdot RT}\right)} = \left.\frac{k_{des} \cdot N_{As*}}{\left(k_{ad} \cdot N_0 \cdot \left(1 + \dfrac{D_{gas}}{\delta \cdot RT \cdot k_{ad} \cdot N_0}\right)\right)}\right|_{p_{H_2}=1\,atm} =$$
$$= \frac{N_{As*}}{K_{ad} \cdot N_0 \left(1 + \dfrac{I_{gas}}{I_{ad}}\right)} \approx \left.\frac{N_{As*}}{K_{ad} \cdot N_0 \left(\dfrac{I_{gas}}{I_{ad}}\right)}\right|_{\frac{I_{gas}}{I_{ad}} \gg 1}. \tag{6}$$

According to equ. (6) the real amount of parasitic dopant pressure is smaller than its equilibrium value that corresponds to the given adsorbate concentration of dopants on the surface. The equilibrium pressure would be established when the dopant gas flow could be suppressed totally. The real parasitic dopant pressure divided by its equilibrium value forms the same fraction as the readsorption flow divided by the gas diffusion flow of dopants.

$$\frac{p^s_{AsH_2}}{(p^s_{AsH_2})^{eq}} = \frac{I_{ad}}{I_{gas}} \tag{7}$$

Taking equ. (1) into account an expression can be formulated that connects the parasitic dopant pressure with the surface dopant concentration of buried layer.

$$p^s_{AsH_2} = \frac{(N_{As_{Si}^+})_{tot}}{\dfrac{K_{inc} \cdot K_{ad}}{F_1}} = \frac{I_{ad}}{I_{gas}} \tag{8}$$

From equ. (8) we find that the parasitic dopant pressure linearly depends on the surface dopant concentration of buried layer, yet its real amount is determined by the fraction formed by dividing the readsorption flow by the gas diffusion flow of dopants.

At low pressure conditions, for instance at $p_{H_2} = 0.1$ atm, the gas flow of dopants is expected to be ten times larger than at atmospheric pressure because of the reciprocal linearity of the gas diffusion coefficient. So, the parasitic dopant pressure should be ten times smaller at a total pressure of 0.1 atm. From the difference of K_{ad} for AsH_2 species and As species (KOKOVIN et al.) it can be deduced that K_{ad} may be increased by about five times in consequence of total pressure reduction to 0.1 atm, however, at least by the same factor k_{ad} must increase and so does I_{ad}.

Buried layers are highly doped and As_2 species may be desorbed from the surface instead of AsH_2 species (CARLSSON et al.). For such conditions equ. (8) delivers us the root of the As_2 partial pressure.

The concentration peak of lateral autodoping

Lateral autodoping takes place by an active doping process which is comparable with intentional layer doping, and it is controlled by the amount of parasitic partial pressure of dopants above the surface. At the beginning of layer growth this partial pressure has its maximum value causing a concentration peak of lateral autodoping. Consequently this concentration peak is located in the interface plane between substrate and epitaxial layer.

$$(N_{\text{lat, max}}) = \frac{A \cdot p^{s}_{AsH_2}}{1 + B \cdot (v_{epi} - v^*_{epi})} = A \cdot p^{s}_{AsH_2}\big|_{v_{epi} \leqq v^*_{epi}} \tag{9}$$

where

$$A = \frac{K_{inc} \cdot K_{ad}}{F_1}, \qquad B = \frac{K_{inc}}{F_1 \cdot N_0 \cdot k_{des}}.$$

Dopant incorporation will be influenced by layer growth rate if the latter is chosen larger than its critical value v^*_{epi}. Below this critical value layer growth rate does not effectuate layer doping, and dopant incorporation is performed as a simple solution process. For dopant pressures high enough equ. (9) should be formulated with respect to As_2 species by applying the root of the dopant partial pressure.

Combining the equs (8) and (9) equipes us with an expression uniform for both AsH_2 species and As_2 species. It relates the lateral autodoping concentration peak to the surface dopant concentration of buried layer:

$$(N_{\text{lat, max}}) = \frac{(N_{As^+_{Si}})_{\text{tot, b.l.}}}{1 + B \cdot (v_{epi} - v^*_{epi})} \cdot \left(\frac{I_{ad}}{I_{gas}}\right) = $$

$$= (N_{As^+_{Si}})_{\text{tot, b.l.}} \cdot \left(\frac{I_{ad}}{I_{gas}}\right)\Bigg|_{v_{epi} \leqq v^*_{epi}} \tag{10}$$

According to equ. (10) minimizing of lateral autodoping peak requires reduction of the surface dopant concentration of buried layer as well as that of readsorption, increase of gas diffusion (yet, if possible, without a simultaneous increase of desorption), and, at last, such conditions that the factor $B \cdot (v_{epi} - v^*_{epi})$ is greater than unity.

References

CARLSSON, J. O., BOMAN, M.: Chem. Vapour Deposition, The Electrochem. Soc. Amer., McD. ROBINSON and others, 1984, p. 129

KOKOVIN, G. A., TESTOVA, N. A., TITOV, A. A., MORGENSTERN, TH., KÜHNE, H.: Cryst. Res. Technol. **20** (1985) 1583

KÜHNE, H.: ibid. **17** (1982) 1097

Kühne, H.: ibid. **21** (1986) 1495
Kühne, H.: ibid **23**, (1988) **327**
Reif, R., Dutton, R. W.: J. Electrochem. Soc. Amer. **128** (1981) 909
Wong, M., Reif, R.: IEE Trans. Electr. Dev. **ED-22** (1985) 83
Wong, M., Reif, R., Srinivasan, G. R.: ibid **ED-22** (1985) 89

(Received, accepted March 10, 1988)

Author's address:

Dr. H. Kühne
Institut für Halbleiterphysik der Akademie der Wissenschaften der DDR
W.-Korsing-Str. 2, PSF 409
Frankfurt/Oder 1200

Cryst. Res. Technol.	23	1988	9	K88–K91

M. NAGABHOOSHANAM

Department of Physics
P.G. College of Science (O.U.)
Saifabad, Hyderabad, India

Etch Pit Morphology in CdTe

1. Introduction

In majority of the semiconductor devices, the dislocation content changes due to (i) ambient temperatures developed (while it is in use) and (ii) free electrons and holes introduced by various defects in the crystal structure. Therefore the transport properties are likely to differ with the constant use of the device. Hence the study of dislocations in particular of semiconductors has a vital importance. Some of the electrical and optical measurements relating to dislocations were reported earlier on CdTe (NAGABHOOSHANAM, HARI BABU; BUCH et al.; FAUST, SAGAR; KROGER; DE NOBEL.). Recently the author (NAGABHOOSHANAM, HARI BABU) also has studied the dislocations in CdTe by developing two different etchants to reveal boat and triangular type etch pits. A correlation between etch pit density and microhardness number was also made. But much emphasis on the shape of dislocation etch pit due to the poison content in the etchant was not made. Therefore, with a view to understand the role of poison on etch pit shape and the temperature effects on dislocation content, the author has undertaken etching, annealing and microhardness studies on single crystals of CdTe.

2. Experimental

CdTe single crystals grown by Bridgman technique were used in the present investigations. To study the dislocation structure and their distribution in these samples four etchants E-1, E-2, E-3, and E-4 are developed. These solutions are obtained by adding 4, 6, 8, and 10 ml of $AgNO_3$ to the basic etchant E [NHO_3(10 ml) + H_2O(20 ml) + + $K_2Cr_2O_7$(4 g)] (INOUE et al.). The time of etching for all the etchants is 120 s. Etched samples were observed under a Universal Research Microscope (Carl Zeiss, Jena).

Vickers indentor attached to the microscope was used to estimate microhardness of the sample. Every time a freshly cleaved surface was used for the investigation. The identations were performed at a load of 100 P and the time of indentation being 10 s. The microhardness value is calculated using the expression.

$$H = \frac{1.8544 \times P}{d^2} \text{ kg} \cdot \text{mm}^{-2}$$

where P is the load applied in grams and d is the length of the diagonal of the identor impression in millimeters. The length of the diagonal was measured using a micrometer. At least ten indentations were performed on each sample and a number of specimens were taken. Microhardness value is an average of all such measurements. The accuracy in the measurement of microhardness is 0.1 kg · mm^{-2}. The samples were annealed at 300 and 450 °C for 6 h using a resistance furnace in vacuum with a controller setup. Etching and microhardness studies on annealed and unannealed samples have been done at room temperature on the cleavage surface determined as (110) from the X-ray diffraction studies.

3. Results and discussion

Figures 1a, b, c, and d show the photomicrographs of CdTe single crystal surfaces etched for 120 s in the etchants E-1, E-2, E-3, and E-4, respectively. The etch pits on matching surfaces etched independently in all the four etchants showed one to one correspondence (micrographs are not shown). This suggests that the etch pits are formed at the emerging points of dislocations. These results thus conclude that all the etchants are capable of revealing dislocations. One can observe in Figure 1a flat triangular pits; in Figure 1b pits with one vertex as curved; in Figure 1c, boat shaped pits and in Figure 1d, elongated boats with some depth. As we approach from

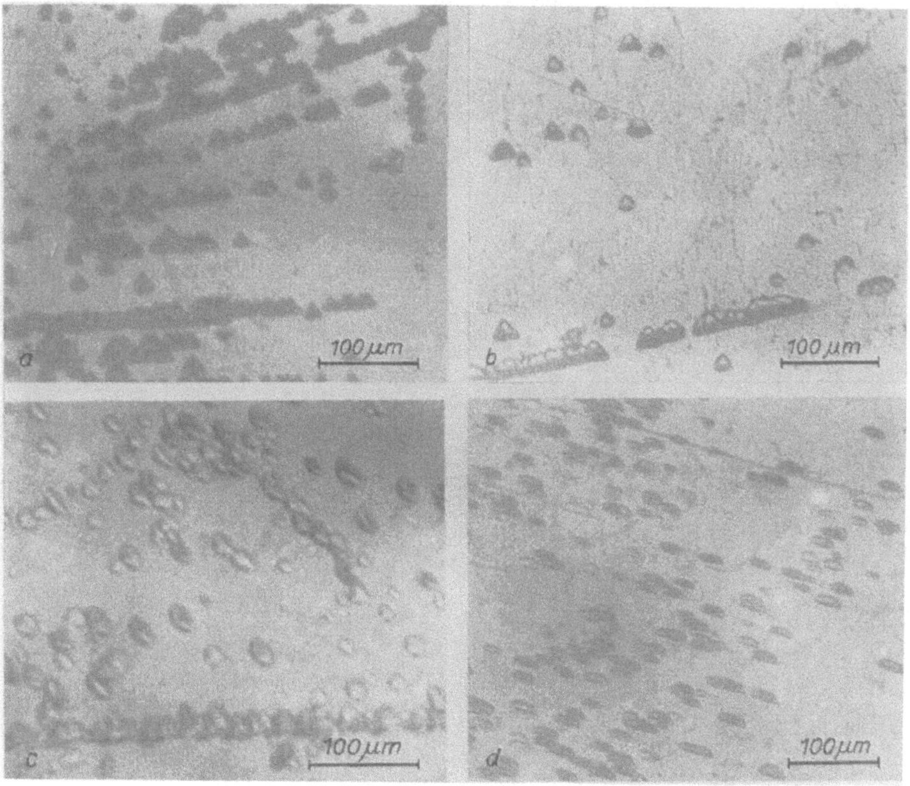

Fig. 1. Photomicrographs showing dislocations on CdTe cleavage surface etched in (a) E-1, (b) E-2, (c) E-3, and (d) E-4

Figures 1a to d there is an improvement in the depth of the pit and clarity in the cleavage steps. This may be due to continuous decrease in surface dissolution rate with the increase in the silver ion concentration in E. Inoue et al. have also suggested that the difference in etch pit shape is due to the difference in dissolution rates of the etchants. It is also seen that the etchant E-1 has equal dissolution capacity in all the three directions of vertices of the pit. But the dissolution rate in one direction decreases more than the other two with the increase in silver ion concentration, a result of which we observe boat shaped pits in E-4. This observation suggests that the silver ions in the etchant may have preferential dissolution rates.

Figures 2a, b, and c represents photomicrographs of etched (110) faces annealed at 300 °C, and annealed at 450 °C CdTe in the etchant E-4, respectively. These micrographs show that (i) dissolution rates of the same etchant on faces of differently annealed samples are different and the rates are unpredictable, (ii) etch pit density increases with the annealing temperature. The reason for (i) may be due to the change in the dislocation core energy as a result of annealing process. The observation (ii) appears like an anomaly. The vickers microhardness numbers (28, 32, and 38 kg · · mm^{-2}, respectively) found on these surfaces also support the observed anomaly. This may be understood as due to the creation of additional Cd- and Sb dislocations

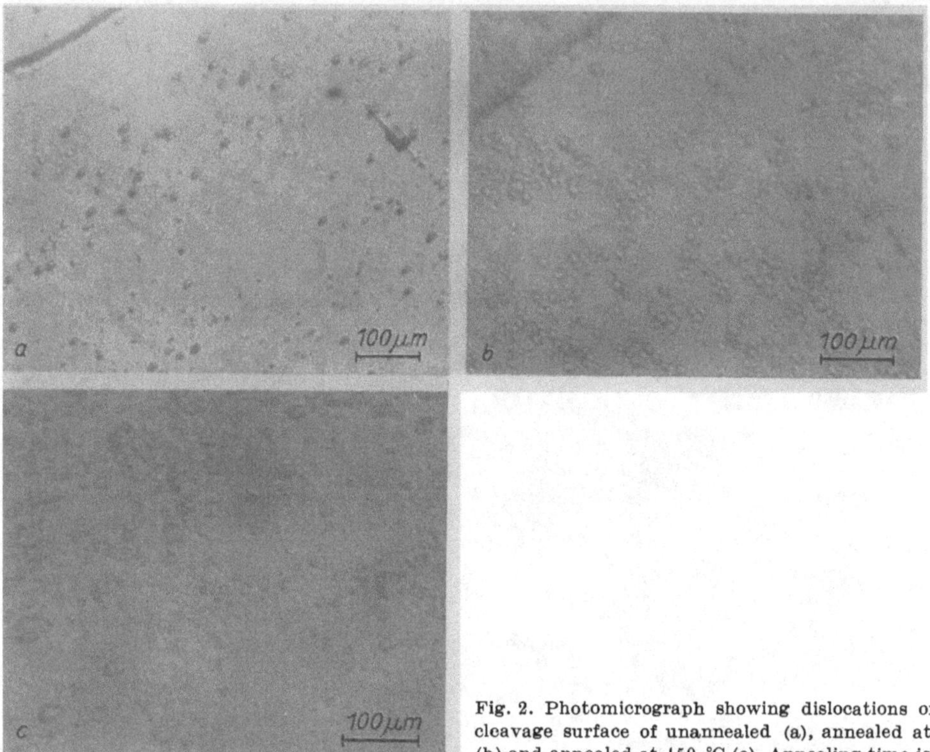

Fig. 2. Photomicrograph showing dislocations on CdTe cleavage surface of unannealed (a), annealed at 300 °C (b) and annealed at 450 °C (c). Annealing time is 6 h

Fig. 3. Photomicrograph showing both boat and triangular etch pit dislocations in chlorine-doped CdTe

created during annealing process. Such dislocation creation can be thought as a result of difference in the vapour pressures of the constituents of CdTe.

Figure 3 shows a typical photomicrograph of an annealed face of CdTe:Cl etched in E-4 showing both boat and triangular pits. These triangular pits may be those sites were there are impurity (chlorine) complexes with their energy matching with the dissolution energy of the etchant. Some such chlorine complexes in CdTe:Cl were also observed earlier (Legros et al.; Agrinskaya et al.).

The author thanks Professor V. Hari Babu, Dept. of Physics, Osmania University for his constant encouragement.

References

Agrinskaya, N. V., Arkadeva, E., Matveev, O. A.: Sov. Phys. Semicond. **4** (1970) 347
Buch, Fredrik, Norman Ahlquist, C.: J. Appl. Physics **45** (1973) 1756
De Nobel, D.: Philips Research Reports **14** (1959) 361
Faust, J. W., Jr. Sagar, A.: J. Appl. Physics **31** (1960) 331
Inoue, Iwaotermoto, Shigetoshi Takayangi: J. Appl. Physics **33** (1962) 2578
Kroger, A., De Nobel, D.: J. Electronics **1** (1955) 190
Legros, R., Marfaing, Y., Triboulet, R.: J. Phys. Chem. Solids **39** (1978) 179
Nagabhooshanam, M., Hari Babu, V.: J. Mat. Sci. **20** (1985) 4329
Nagabhooshanam, M., Hari Babu, V.: Cryst. Res. Technol. **19** (1984) 643

(Received, accepted April 14, 1988)

Author's address:

Dr. M. Nagabhooshanam
Department of Physics
P.G. College of Science
Saifabad
Hyderabad — 500 004
India

Crystal Research and Technology

Zeitschrift für experimentelle und technische Kristallographie

Edited by H. Neels, Leipzig

Papers in Number 6, 1988

Original Papers

AKADEMIE-VERLAG · BERLIN

optimization

A journal of mathematical programming and operations research

Issued at the Ilmenau Institute of Technology by K.-H. Elster, assisted by an international Editorial Board

The journal optimization addresses readers dealing with mathematical programming, operations research and stochastic processes. The journal publishes carefully selected theoretical and applied papers in English, Russian, German or French, covering the following fields: linear, non-linear, stochastic, parametric, discrete and dynamic programming; control theory, theory of games, mathematical modelling of economic and technological processes, inventory theory, queueing theory, reliability theory, theory of renewal, network methods and related problems.
The contributions contain recent results regarding theoretical statements, numerical methods or applicational aspects of the above given topics. Furthermore, surveys, reports on computer routines, reviews of books as well as information on conferences are published.
The journal ist aimed at mathematicians and mathematically interested persons researching, teaching or applying methods of mathematical programming, operations research, and stochastic processes at universities, in industry, planning and management as well as in computing centers.
A forum is offered by the jorunal to scientists engaged in international research.

Publication dates:

The journal is issued annually in one volume of 6 numbers.
It was founded in 1970 as "Mathematische Operationsforschung und Statistik". Volume 19 will be issued in 1988.
Number of pages: 160.
Size: 17 cm × 24 cm. Annual index.
Annual subscription price: 198,— DM; Price per number: 33,— DM.
Order number: 1093. ISSN 0233-1934.

Possibilities for subscription:

Orders are to be addressed
In the GDR: to the Central Post Office for the sale of newspapers by indicating the postal code of the orderer or, to the publishers:
Akademie-Verlag Berlin, Leipziger Straße 3—4, PP 1233, Berlin, DDR 1086.
In European countries: to "Kunst und Wissen", Erich Bieber GmbH, General Wille-Str. 4, CH 8002 Zürich

AKADEMIE VERLAG BERLIN
Deutsche Demokratische Republik
DDR-1086 Berlin, Leipziger Strasse 3-4 Postfach 1233

Continued from cover page 4

Short Notes

Book Reviews

Crystal Research and Technology
Volume 23 1988 Number 5

Contents

Original Papers

Continued on cover page 3

Crystal Research and Technology is indexed in Current Contents/PC & ES.

FSC
www.fsc.org
MIX
Papier | Fördert gute Waldnutzung
FSC® C083411

Zeitfracht Medien GmbH
Ferdinand-Jühlke-Straße 7
99095 Erfurt, Deutschland
produktsicherheit@kolibri360.de